Arthur Kirk

The Quarryman and Contractor's Guide

How to Remove Rock at Least Cost

Arthur Kirk

The Quarryman and Contractor's Guide
How to Remove Rock at Least Cost

ISBN/EAN: 9783743383845

Manufactured in Europe, USA, Canada, Australia, Japa

Cover: Foto ©berggeist007 / pixelio.de

Manufactured and distributed by brebook publishing software (www.brebook.com)

Arthur Kirk

The Quarryman and Contractor's Guide

.

THE

Quarryman and Contractor's

GUIDE,

— OR —

How to Remove Rock at Least Cost,

INCLUDING ARTICLES ON

The importance of Blasting Explosives; Wrong Treatment of Explosives; Transportation and Rate of Freight paid on Explosives
Great Loss sustained by prohibitory regulations against Explosives by R. Rs.; Rate of Insurance on Explosives; How to
build Powder Store-houses or Magazines; Cause of Accidents by Explosives; Rules relating to Handling and Storing;
How to thaw frozen Dynamite safely; Bomb Proof Shelters for Quarrymen; Waste of money by working
Quarries on any old plan; Contrast of Hand and Power Drill plan of making holes in rock; Contrast between large and small Blasts; Stripping rock of dirt by Hydraulic Washing by means
of a force pump; Nature of Dynamite as an Explosive; How to use Dynamite and Blasting Powder, Details of Electric Blasting; Detailed account of Blasting at Johnstown; Stump Blasting; Rock Drills; Air Compressors ; Coal Augers, etc., etc.

— BY —

✧ ARTHUR KIRK, ✧

OF PITTSBURGH, U. S. OF AMERICA.

INTRODUCTION.

IN presenting to the public the following book on Blasting Explosives, Rock Drills, etc., we intend to insert much new matter which we expect will be both interesting and profitable to everyone connected with **Railroad Contracting, Quarrying, Tunnelling, Mining, Shaft Sinking, Stump Blasting** and all kindred subjects, and we ask your careful inspection of it, as we feel assured that before you get through with it you will find **something** that will repay for reading much you already know. First, we ask your attention to

THE IMPORTANCE OF BLASTING EXPLOSIVES.

As now used, explosives are of far more importance among the comforts of life than most persons have any knowledge of. Without explosives we could not make iron or steel, or make a single piece of glass, or build a substantial building of any kind, or a railroad or canal; and without powder all our fine-drawn theories about coast defences or military organizations, drills and equipments would be pure nonsense. Let the reader stop a moment and try to imagine a state of society **without explosives** to get out coal, iron ore, limestone and building stone. This, of course, would carry with it the total annihilation of all iron and steel tools or machines; our saws, hatchets and axes would at once disappear and we would step away back to the stone age. There could be no substantial buildings built without powder to blast stone for the foundation and the greater part of all red brick are now made from blasted material. None of our fine block-stone pavements could be gotten out without blasting explosives, and blasting has to be used to open the quarries where our fine stone trimmings for brick fronts are gotten.

True, the ancient Egyptians appear to have gotten out and handled very large blocks of stone without powder, but that is now a lost art, for which great search has been made without success.

The very pleasant practice of taking an annual vacation for pleasure and healthy recreation would have to be abandoned. There could be no buggies, carriages or street cars to take us to the station or steamboat landing, and there could be no rushing railroad train to take us at the rate of thirty-five miles per hour to the cool, inviting, health-giving summer resort, and as for steamboat travel or a trip across the ocean in six, seven or eight days, it would be utterly impossible, for without blasting to get out iron ore, coal and limestone, there could not be a single one of our fine-floating and swift-rushing floating palaces, for the best floating craft that could then be had would be a burned-out canoe which we might paddle if we could find a suitable long splinter from a lightning-riven tree to use as a paddle.

And there could not be any nice basket of provisions to accompany us for even one day off to the woods, with its fine assortment of pies, cakes or cookies, for without blasting there could be no iron cooking stoves or ranges to cook these on, nor any extra fine roller process flour to make them from. But the best grinding or cooking that could be had would be two women grinding at the mill, and then mixing it with water, and without bolting or salting it, working it into cakes and baking it on the coals, fresh for every meal, and our fair lady friends couldn't have a nice new dress (just made for the occasion), with latest patterned towers of very fancy colored trimmings, for without blasting there could be no iron or steel to make the carding, spinning or weaving machinery to make the fancy dress goods, nor a pair of scissors to cut them, nor a needle to sew them, or even a pin to hold them in position; in short, without the blasting of minerals we would have to abandon all that steam engines now help to produce; all our comfortable home furniture, kitchen, bedroom and parlor sets; carpets would be unknown and we would step at once into the simple style of living, practically as Adam and Eve did 'when they first started housekeeping outside the gate of the Garden of Eden.

Having thus called your attention to the importance of explosives, I ask why should such an important factor in every-day life be subjected to the

WRONG TREATMENT IMPOSED ON EXPLOSIVES.

I see no reason why explosives should be charged such high rates of freight on railroads as double first-class rates, for all railroad managers should know that the more explosives used along their railroad the better it is for the road. Because it not only produces large amounts of freight in the shape of coal, stone, etc., but it calls there a great increase in the population who are every day paying large sums

FIG. 1.

THIS IS THE MAN WHO LIVES IN A HOUSE MADE WITHOUT EXPLOSIVES.

THIS CUT IS KINDLY LOANED BY M'GRAW BROS., MAKERS AND DEALERS IN ROOFING MATERIAL, 64 FEDERAL STREET, ALLEGHENY, PA.

as passage fare which railroads, running through a purely agricultural district, would be glad to have. Explosives are the very life of a railroad, **almost as necessary as water for their locomotives,** and to levy such a high tax rate on explosives is almost as foolish as to tax locomotives for the water they use, and **why should powder not be carried every day** on railroads on the same rules as **dry goods?** No reason can be given for restricting its cartage to one or two days each week, or

FOUR DAYS IN A MONTH AS SOME RAILROADS DO.

But many reasons can be given why it should be carried every day, and

First, I would say no quarryman or railroad contractor or coal pitt manager can tell how much powder he will need during next week, because a quarryman or railroad contractor may have enough powder to do him in the ordinary way until next powder day. But he suddenly comes to a seam where

4

he can use twenty, thirty or forty kegs in one blast to good advantage, and then he is out of powder for eight or ten days, and must he stop perhaps **fifty men** and throw them idle if he waits for powder day? This he will not do on any account ; but inside of twenty-four hours he has plenty of powder, in defiance of railroad rules. But how does he do this? I do not think it my business to tell all the plans of getting it over the railroad ; but to prove that railroad men cannot prevent explosives being smuggled over a railroad, I mention one plan · I have known coarse quarry powder emptied out of kegs into new two-bushel bags and shipped as **beans,** and again have seen a flour sack (which I knew contained a keg of powder) standing on a very cold winter day within ten feet of a red-hot stove in a smoking car, and within three feet of the superintendent of the same road, who calmly smoked his cigar unconscious of danger, but who, in spite of all arguments to the contrary, would only permit powder to be carried over the railroad one day every two weeks.

Second. Our second reason is, much local **business is done over railroads by country storekeepers** and country peddlers, who collect farm produce and haul it ten to fifteen miles to a railroad station and then ship it to the city, leaving their teams to rest until they go to the city and return the next day with many goods for their customers, among which may be an order for more or less powder; but he is told that "Yesterday was powder day on your railroad and your powder cannot be shipped as such until yesterday a week." He does not hesitate a moment. He knows he must have the powder with him and some plan is adopted by which it goes over the road packed in a barrel of rice or in a box as soap ; and the railroad loses the high rate freight on it as powder, and it serves them right, because railroads have no right to make rules that cannot be obeyed.

Third, and last reason why explosives should be carrried every day whenever offered, is that railroad officers know as well as we do that the smuggling of explosives has been carried on over their railroads for the last twenty years, at least, as above described, without doing any damage to anything and with great danger to both passengers and railroads, and yet for some reason best known to themselves, they still insist on **"our rule,"** to the manifest inconvenience of their patrons and the increased danger to passengers by having it smuggled on passenger cars, and lose freight on it when shipped openly every day as explosives.

But I will close my remarks on this subject by a quotation from Col. Majendie's report for the year 1884. Col. V D. Majendie is the Chief Inspector of Explosives for Great Britain, and as such has for years done nothing but make everything connected with explosives his special study, and he is, I think, the best informed man in the world on such matters. He says as follows :

This prohibitory policy, we desire emphatically to repeat, supplies a temptation to the surreptitious conveyance of the explosives which it is certain is not always resisted. As we stated last year, the practical effect is thus to introduce a greater risk than that against which the policy of the railway companies is professedly directed. If it is dangerous to carry dynamite openly in a properly constructed van by good trains, separate from all dangerous articles and subject to the various precautions imposed by the by-laws, it must surely be more dangerous to convey it, concealed perhaps in unsuitable parcels under the seat in a smoking car or among the miscellaneous baggage of a passenger train. The railway companies must know as well as we do that there is reason to believe that such surreptitious conveyance is deliberately practised on their lines, and upon what grounds they justify the continuance of a policy which tends directly to the encouragement of that practice, we are at a loss to imagine.

COL. V. D. MAJENDIE,

Her Majesty's Inspector of Explosives for Great Britain.

WORSE THAN DYNAMITE.

The Lucifer Match and Coal Oil Do More Damage.

The Detroit *News* takes this hopeful, philosophic view of dynamite: "With all the powers of dynamite and all the attempts to use it, the results effected are comparatively insignificant. It is true a czar has been killed with it, but hundreds of kings have been killed without it. The lucifer match, of which every household contains hundreds, is much more dangerous, and when first invented inspired nearly as much terror. It destroys more property and more lives than dynamite does. It has in it the power and potency of infinite mischief, if there were men enough wicked enough to use it as it might be used. The rascally fools who exploded a dynamite bomb under the walls of their employer's shop in New York, with but little injury, might have laid the whole establishment in ashes with a single lucifer match properly placed. The revolver, the shooting walking-stick, the slung-shot, the policeman's club, are all more dangerous weapons of offense, and number their annual victims by the score to every single victim of dynamite or nitro-glycerine; while the numerous deadly poisons, which can be administered without detection, afford opportunity of doing a hundred times the mischief, with safety to the perpetrator, that dynamite can do.

"If it be considered from the point of view of the probability of accident it presents, it sinks into insignificance when compared with the thousand dangers through which we calmly and thoughtlessly walk every day, and which we despise because we are familiar with them. Not to mention the hundreds of contagious and infectious diseases, to which we expose ourselves without knowing it, or knowing it, trusting like a Turk to Kismet to save us, it would require more space than the *News* has at command to enumerate the exploding kerosene lamp, the oil stove, the open elevator, the bursting boilers, the burning theaters and hotels, the hot flue, the colliding and the derailed train, and the hundreds and thousands of inventions for modern convenience and luxury, which conceal deadly dangers that frequently break loose and destroy hecatombs of human beings. We venture to say that fried beefsteak and American pie have destroyed more of the inhabitants in the past twelve months than dynamite can be charged with in all its history all the world over; while the victims of nervous diseases, brought on by over-indulgence in strong tea and coffee, among those who regard the more wholesome beer and ale with pious aversion, would furnish regiments where the dynamite victims would not furnish corporals' squads. But we have got used to all these things, and they no longer terrify us." And I may add the following:

KITCHEN RANGE BOILER EXPLODES AND CAUSES DEATH.

A boiler attached to a kitchen range of Frank T. Sherwood, of Hunter's Point, L. I., exploded with terrific force recently, instantly killing Charles M. Sherwood, aged five years, and fatally injuring Kellogg Sherwood and burning his mother terribly. She will probably lose the sight of both eyes. Everything in the room was destroyed.

Some ten years ago a **terrific sewer gas explosion** alarmed the residents on Penn avenue from Twenty-eighth street to Twenty-ninth street, and was heard ten blocks away. November 6, 1884, Robsonia (Pa.) furnace fell, killing seven and wounding eight men. Syracuse, N. Y., January 6, 1884, an ash factory exploded and killed two and wounded eight men.

GREAT NEED OF BETTER LAWS RELATING TO EXPLOSIVES.

The great importance of explosives, and also the great danger from their careless handling, calls loudly for better laws relating to their manufacture, shipping and use. Just a few days ago the quiet town of Washington, Pa., was shaken as by an earthquake by the explosion of one hundred pounds of nitro-glycerine. Every person was alarmed, much property destroyed, and one man and two horses killed; a great many lives were endangered by a fool-hardy, useless risk in transporting nitro-glycerine through a town, and I see no reason why there should not be a law to protect lives and property against such damage. A suitable law on this subject would be a great protection to everyone. It would be a protection to owners of works, and to everyone engaged in them, and to shippers, haulers and users of explosives, and every person in the community, by throwing every possible precaution around the making, care-taking, transportation and use of explosives. As it is now, there is practically no law on this subject. True, there are some local laws, but all I know of were enacted long ago, and are so unsuitable and far behind the age that they are of no use and are never enforced.

OF INSURANCE ON EXPLOSIVES.

Why should explosives be subjected to so many antiquated and obsolete rules relating to how powder shall be handled and stored—rules that have been disregarded for fifty years and no harm has resulted, showing the utter worthlessness of such rules? And yet their existence in one place is made the excuse for enacting the same rules elsewhere; for instance, the rule with some insurance companies that the insured shall not keep more than twenty-five pounds of powder in store at one time, and the same is enacted into a law in other places.

Now any dealer in explosives who accepts an insurance policy with this clause in it simply puts his head into a noose and practically gets no insurance at all, for almost any dealer in powder now requires to have at least a keg of fine and a keg of coarse rifle powder, and one keg of coarse and a keg of fine blast powder, all open for retailing from at the same time, with a keg or two in reserve ready to open when needed, and therefore we recommend that no dealer in powder accept a fire insurance policy without permission to keep two hundred and fifty pounds of powder; and that he can easily have at the same rate, for if twenty-five pounds of powder should ever explode it would knock the whole house to smash, and what more could two hundred and fifty pounds do? But who ever heard of a fire caused by powder in a store? I have never known or heard of one and don't think there ever was one.

RELATING TO POWDER MAGAZINES.

This is another subject on which overwise ignorance prevails. From their great importance to society, as has already been shown, a place of storage or magazine is an absolute necessity, and it should be located as near shipping or using points as possible, without being compelled to haul long distances through populous places; this only increases the danger by extending it all along the line of haul. Yes, but, says the objector, a magazine might explode.   Well, if it does explode, as magazines are now built of walls of wood six inches thick and covered with corrugated sheet iron, the resistance is so slight that an explosion will do but little harm, all going upward—nothing like the damage that has been done and may be repeated every day by steam boilers exploding, as witness the explosion of Rees, Graff & Co.'s mill, near Thirty-third street, some twenty years ago, when fragments of the boiler weighing nearly one thousand pounds were thrown nearly a thousand feet, with all the force of a cannon ball.   This was followed a few years later by the explosion of Zug's rolling mill, when large fragments of the boilers were seen dropping in the river over a thousand feet from starting point, and Groetzinger's tannery in 1889, which killed four men and caused widespread damage.   Thus if everything liable to explode must be kept at a great distance from human habitation, it would require the whole county of Allegheny to contain the city of Pittsburgh alone.

FIG. 2.

BULLET AND FIRE-PROOF RETAIL MAGAZINE,

8x12 and 8 feet high to square, on stone foundation walls two feet high above ground.   Door in middle of side, hung on friction rollers.   Sides and roof covered with corrugated sheet iron.

The above cut shows a modern-built powder magazine as now built with stone foundation two feet above ground to prevent the accumulation of leaves or burnable material under it.   Walls and door bullet-proof, being six inches thick, and a light roof of corrugated sheet iron, so that if an explosion should ever take place the roof will just blow off and explosion go upwards and do but little harm.

8

RULES TO BE OBSERVED

By everyone handling Powder or any other Explosive Substance

1st. Always keep up in your mind a safe fear, and remember it is dangerous, and handle with proper care.

2d. Never permit any explosive material to be taken into a **BLACKSMITH SHOP.**

3d. All explosive material should be kept (when not in use) in a suitable dry place, at a safe distance apart from other buildings, and at all times under lock and key, and only one person in charge of it, and no person but one keeper permitted to enter it.

4th. No lights, matches or fire arms, or means of making a light should ever be taken near where powder is stored.

5th. In making floor of powder house, no nails should be permitted to show above boards, but all secret nailed. Floor should be kept perfectly clean, and no sand or grit permitted to remain on the floor.

OF ACCIDENTS BY EXPLOSIVES.

For twenty years I have made a practice of cutting out of the newspapers every account published of accidents by explosives, and from a casual reading over of these clippings I came to the conclusion that **blunders** would more appropriately describe the great majority of cases of damage by explosives; for instance, the first of these cases that meets the eye reports the death of two men in Kentucky, while drilling out a hang-fire shot. This is always attended with great danger when fired by ordinary fuse. Shots have been known to hang fire twenty-four hours and then explode, to the great danger of every living being within range. This could never occur if the hole had been charged to **fire by electricity.** (See page 30.) Another of these clippings records the destruction of five thousand dollars' worth of property because a lot of dynamite had been stored in an engine room. I have again and again warned our customers against permitting **explosives to enter engine rooms or blacksmith shops,** and yet only last summer I found forty kegs of powder, five hundred pounds of dynamite and ten boxes of caps, all within fifteen feet of the anvil where the blacksmith was making a weld at the time.

Another great cause of accidents with explosives is

THAWING DYNAMITE IN AN IMPROPER WAY.

Dynamite freezes at 42° and while frozen is practically unexplodable by any ordinary means. Some experiments are said to have been made in Germany which exploded frozen dynamite, but practically in this country every winter brings a crop of fatal accidents while attempting to thaw it **in an improper manner.** Most of these occur in blacksmith shops, and I repeat what I have said, that no explosive should ever be permitted inside a **BLACKSMITH SHOP,** as that is the worst place to thaw frozen dynamite that I know of. A very safe plan about large works is to have a frost-proof room, with means of keeping it heated to 70° or 80° and have the dynamite placed around on shelves where it can thaw slowly through and through, and about summer heat maintained.

PLAN FOR DYNAMITE THAWING ROOM.

Below I show a plan (fig. 3), and inside elevation (fig. 4, p. 11), and front elevation (fig. 5, p.12), for such a house, which can be built very cheaply, and on the principle that "an ounce of prevention is worth more than a pound of cure," I recommend every quarry owner to erect one on this plan at once, larger or smaller, as his business may require. I recommend it to be built in the quarry dump, as convenient to the work as possible, and so low down in the ground that it can easily be covered three or more feet deep and all around three sides with the refuse of the quarry (*which has to be moved at any rate*), and with its door end pointed away from the blasting. If thus constructed, blasting can be done quite close to it without endangering its contents.

This house can be costructed of any size to suit the work. The size here given is five feet four inches by eight feet long and nearly eight feet high, so that sixteen-foot stuff will cut to advantage without waste, and is to be provided with five shelves on each side (see fig. 4), ten shelves, and each shelf holding four boxes, will be forty boxes this house is capable of holding. This can all easily be kept warm enough to thaw dynamite. A short coil of small steam pipe, or if that cannot be had, then a small "Summer Queen" stove with double 4-inch wick, placed on floor at A (see fig. 3), and boring three 1-inch auger holes through the door at B (see fig. 5), to let in air, and three same sized holes at D to let vapors out ; the space inside can easily be kept heated to 80° Fahrenheit in coldest weather, and at this temperature in such a chamber dynamite will keep thawed for forty-eight hours and ready for use.

It will be observed the front (fig. 5) is built up of plank, laid flat ; this I recommend should be not less than six inches thick, and the door same thickness.

The reason for this is to prevent it being exploded by any person firing a heavy musket ball at it from a safe distance for the shooter.

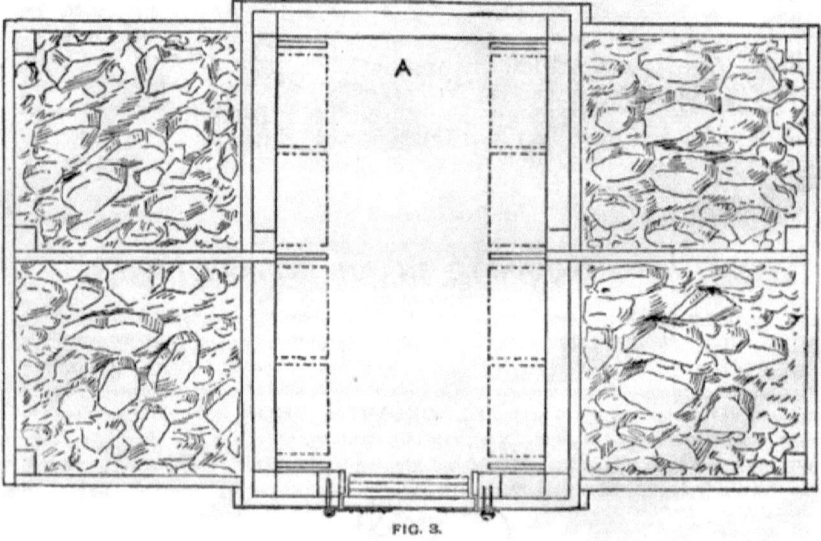

FIG. 3.

10

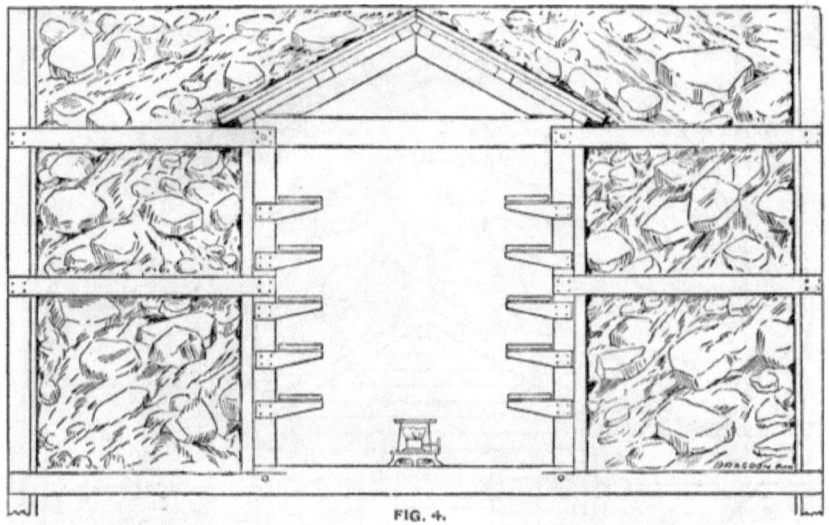

FIG. 4.

ELEVATION OF THAWING HOUSE.

Some readers may think I attach too much importance to **thawing dynamite, preventing accidents, etc.** Some, no doubt, will be ready to say, I have used dynamite (thus and so) for ten years and never had an accident, and it is all nonsense going to such expense, etc. To this I would reply : True, you may have done so and so, and thus far have escaped damage, but you perhaps were not aware how near the danger line you have come a hundred times. You may the very next time pass over the danger line and be torn to fragments, and as I think it wiser to be one mile away than merely escape by a hair's breadth, I recommend and urge everyone to take the safest plan in everything connected with blasting. I once thawed dynamite on hot plates covered with sand level with the floor. I was absent longer than I expected, and when I returned and standing astride of ten pounds in this sand, it suddenly took fire and all burned up, and it was a wonder it did not explode and tear me to fragments. I never ventured to thaw on heated plates again.

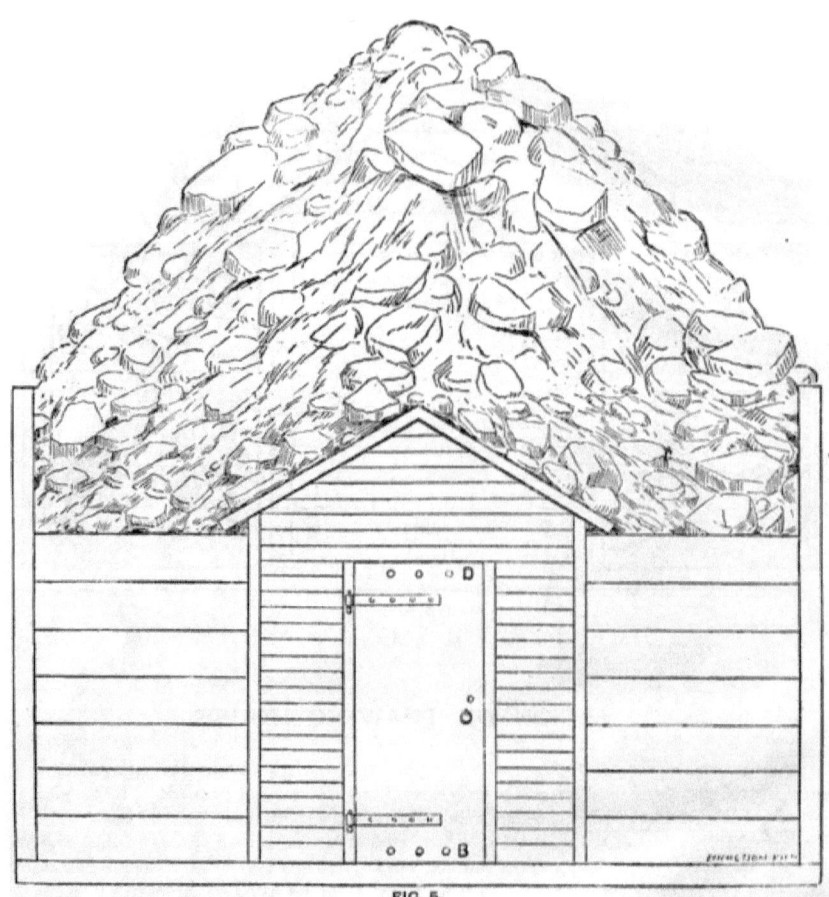

FIG. 5.

FRONT OF THAWING HOUSE AND BOMB PROOF SHELTERS FOR MEN.

Can easily and cheaply be thus made in the dump of every quarry by simply setting up strong timbers where a dump is to be made and covering them with planks, boards or even round poles, and then dumping refuse of the quarry over it ; this will, in a short time, provide bomb proofs where all the men of the quarry can feel perfectly safe from flying rocks.   Only about one year ago an estimable young man, Mr. Mananing, was killed in a quarry near Williamsburg while running away from a blast, and this, too, after working all his life about quarries ; and many lives have been lost in the same way which might have been saved if the quarry had been provided with bomb proofs as above.   And thus I might go on pointing out many cases of damage by explosives, but I think I have shown enough to satisfy any one that all accidents from explosives are preventable by simply remembering that all matter in the world is under certain fixed laws and if we obey the laws of matter in any form there is no danger with it.   For instance, every person knows the laws of matter in the form of water and oil, and, therefore, no sane person would take oil to extinguish a fire, neither would he take water to kindle a fire, and I, therefore, lay it down as a rule that every one handling explosives should keep up in his mind a healthy fear that it is dangerous and treat all explosives with proper care.   I next call your attention to

FIG. 6.

FIG. 7.

The next best plan for thawing dynamite is by means of a thawing kettle shown in figures 6 and 7, which consists of an outside kettle capable of holding twelve gallons of water (see B, fig. 7), and having four iron feet to hold it firm above the fire to be made under and around the kettle.   It also has an inside kettle (A A, fig. 7), which fits inside the large kettle, so that it has a water jacket of two inches around and between the inside and outside kettle sides, and five inches of water between bottoms of inside and outside kettles ; the inside kettle also has an open space (B, fig. 7) six inches in diameter at bottom and an outside and inside shell which forms the dynamite chamber (A A, fig. 7), into which can be placed forty-two sticks of 1¼x8 inch dynamite, and then all is covered by a tight-fitting lid except the small pipe (see figs. 6 and 7), which permits the steam to blow off freely from the inside chamber.   I have now sold these kettles for five years and have never known of an accident with one of them, and every user praises them.

WASTE OF FORCE (MONEY) IN HANDLING MINERALS.

It makes the feelings of the writer fairly rasp to see or hear of men still drilling holes in rock by hand with three men in a gang, one holding and two striking, and these three men thinking they have done extra well if they drill ten feet of hole in limestone rock per day, when these same three men at the same wages, if supplied with a steam boiler and rock drill, would easily average sixty feet of hole per day. If we count that those three men get each $1.50 per day, it makes $4.50 for ten feet of hole or 45 cents per foot, while if these three men are provided with proper tools and drill sixty feet for $4.50, it makes it cost only 7½ cents per foot. Now, if we count twenty-four working days per month, and sixty feet of hole per day, at 45 cents per foot, it will cost $648 to pay for that amount of drilling by hand. But if we furnish a steam drill and pay three men 7½ cents per foot it will only cost $108 ; this from $648 leaves $540, and if we still take off the $40 for coal, oil, etc., we still have a net saving of $500 per month. Thus counting ten working months in a year, makes a clean saving by steam drilling alone of $5,000 a year. And this is only a part of the advantages to be had by using a steam drill. Another advantage is

PLENTY OF HOLES READY FOR BLASTING.

I never visit a quarry where the drilling is done by hand but I find some part of the men working at a disadvantage until the drillers get a hole finished. On the other hand we know several quarries using steam drills where they have a ten-foot hole here, a twelve-foot hole there, a twenty-foot hole yonder, and so on, till they generally have two hundred feet of hole drilled ahead of the blasters, and ready to be fired the moment the loose stone is all taken away from before them. Every experienced quarryman knows what a great advantage this is and must add largely to the yearly profits. And yet another advantage of power over hand drilling is, power-made holes are perfectly round, while hand-made holes are always triangular, as per figures 8 and 9.

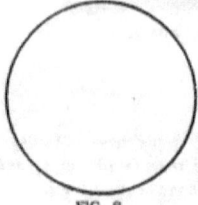

FIG. 8.

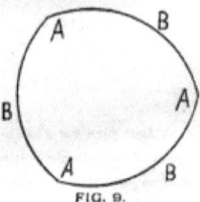

FIG. 9.

Figure 8 represents a machine-made hole, perfectly round, and when explosives are fired in such a hole, the force of the blast being equal all around, it is held in longer, until its explosive gases are fully developed, and when the blast finally bursts, it does far more execution than when fired in a hand-made hole. Figure 9, representing a hand-made, and, of course, a three-cornered hole, if made by the ordinary chisel bit, where the corners (A A A) have to withstand the concentrated pressure on the walls (B B B), the corners (A A A) give way and the explosive gases escape before their full force is developed, and in underground work this fills the place with smoke, which is *imperfectly consumed powder*. Another advantage of power-made holes over hand-made holes is

A LARGER HOLE AT BOTTOM.

Hand drills are made thin so as to cut quickly, but they at same time wear away faster than steam drill steels, and when you add to that their triangular shape, the powder space at bottom of hole must be very small, hence it must be squibbed and squibbed, again and again, to enlarge the hole.

THE GREAT DANGER IN SQUIBBING HOLES.

This is always attended with great danger to the blaster, especially if done with fuse, and more so if done with cotton fuse. By referring to the cuttings from newspapers, already referred to in this book, I find that almost every month some one is injured by a premature explosion while squibbing a hole. For this there are several reasons and first I would mention as far as I have ever seen : Squibbing is almost always done in a hurry; the rock has to be cleared away before the drillers can start to hammer down a hole ; then, if it takes a day to put down a ten-foot hole by hand, before the hole is down all hands are idle around it and then, "Hurry and squib that hole," is the order, and without waiting to let any remains of fire from first squib die out, a fresh charge is poured in, and after the hole is nearly charged again with several kegs of powder it reaches a smoldering fire in some piece of fuse that has been forced into a crevice in the rock which, on account of a limited supply of air in the hole, burns very slowly and holds fire much longer than same fuse would do in the open air, but still contains enough fire to ignite the blast, and off goes the charge and another man or more is killed, all because **that hole must be squibbed in a hurry** and cotton fuse used because it is cheap. If holes must be squibbed, it is much cheaper and safer to fire the squib by electricity, for the wires used instead of fuse cannot hold fire one instant after the charge has been fired, and it is done much quicker, and at the will of the blaster, when every person near is safe from flying stones **of squibbing holes.***

DOES SQUIBBING HOLES PAY ?

On this subject I have found a great diversity of opinion—and that, too, among blasters of great experience ; some insisting that by far the best plan is to spring every hole before blasting, and just as many insisting that a great deal of time and powder is lost while springing holes.

I think this great diversity of opinion may be accounted for by taking into account the different experiences of the different men. For instance, every blaster of much experience must know that the same kind of stone blasts differ in different quarries, and a man having worked for years in a quarry that worked well by squibbing, naturally falls into the idea that squibbing must be used in every quarry. And yet another reason may be the man in favor of always squibbing, has only had experience with hand-made holes, which are both slow and expensive, and of course, when once down, must produce a great amount of stone to pay for the drilling, as has been already shown in this book (page 14). A hand-made hole ten feet deep and taking three men a whole day to hammer it down, costs $4.50, or 45 cents

*Squibbing a hole in quarry work means enlarging it by means of a repetition of small blasts to make a pocket for the large blasts.

per foot. While at the same time three men with a good steam drill can drill sixty feet of hole per day, at a cost of 7½ cents per foot, or three men at $1.50 each per day or $4.50, at this rate the steam drill will have six holes each ten feet deep and if these holes are placed, say ten feet apart, and charged with mixed charges, of say one stick of dynamite with electric exploder in it at the bottom, then one quart of black powder, then another stick of dynamite, then another quart of black powder, then another stick of dynamite until the hole is filled up five to six feet, or to within two feet of the top in the usual way and all fired with a battery as hereafter described, after firing it thus it will be found that at least twice as much stone has been loosened than could have been done with the same amount of explosives used in squibbing and **blasting with a single hole at a time.**

I have never found much advantage in squibbing power-made holes—and another advantage of power-made holes is being able to make large round holes cheap and near each other and use electric blasting, for which see page 30.

MAKING SMALL BLASTS

Is another cause of great loss to contractors, quarrymen, etc. We have seen many quarrymen who rejoice when they get off a shot loosening ten ton, and when they loosen up twenty ton, talk about it for weeks as something wonderful. By way of comparison, we call your attention to the following extract from a letter, giving account of work done by a steam drill :

HEAVY BLASTING.

"Perhaps the largest blast ever made in the actual operation of stone quarrying for the market occurred at quarry No. 4 of our company, at Hilltown, Pa., on Friday, the 4th of last month. We drilled one hundred and thirty-one holes in two parallel lines, the holes alternating and about seven feet apart, each eight feet deep, with one of your No. 'F' Eclipse Drills. The time occupied in drilling was thirteen days, and took two men (one to operate the drill, and one on the boiler). These holes were then charged with powder and dynamite, fifteen kegs of the former and fifty pounds of the latter. After the charges were prepared, the whole was exploded simultaneously by electricity ; between 12,000 and 13,000 tons of limestone (by actual measurement) were displaced by this blast, opening a seam through which a team could easily be driven. This work was done at an expense of $94.60 and a saving to the company of over five hundred dollars. We have tried steam drills of several different makes, and unhesitatingly pronounce yours by far the best, and commend it to all in our business. I am, very respectfully yours,   JOHN A. LOGAN, Jr.,

    CARBON LIMESTONE CO.,   *General Manager.*"
        YOUNGSTOWN, OHIO.

This is probably the most effective blast on record, showing thirty tons of limestone removed to one pound of explosive.

MONEY LOST IN LOADING STONE FOR SHIPMENT.

Another way money is lost in quarries and railroad work, is loading it into a wheelbarrow as if it was eggs, for fear of breaking the wheelbarrow and having to place every stone carefully so as to avoid upsetting the wheelbarrow, and then take a man at $1.25 per day to run that wheelbarrow one hundred feet and then up a steep grade to the top of the railroad car. Quarrymen have told me over and over again they would rather sledge a wheelbarrow load than wheel it as above.

Then another objectionable way is to load it into carts and have a horse or mule to haul it. This is still too expensive, because every man loading a cart has to lift by his own strength every piece of stone about five feet from the ground to put it into a cart, and after the cart is loaded a man must go with every ton and walk away off with the horse. Another less expensive, but still objectionable way is to have small railroad tracks on which to run cars hauled by a horse. This is also objectionable because the rails are liable to be broken every shot and a horse is constantly needing shoeing or harness repaired, fed, watered, etc. But the most economical is wire rope hoist and conveying, of which more will be said hereafter. After calling the attention of our patrons to the above leaks, where many thousands of dollars are lost every year in quarry work for want of a little knowledge, I am was tempted to repeat the old story of a darky who had been engaged to kill a calf. After the job done he was asked : "Well, Sam, how much must I pay you for killing the calf?" He promptly replied, " Two dollars and a half, Massa." "Why, how is that, Sam, the whole calf is not worth over two dollars and a half, how then is it worth two dollars and a half for killing it?" " Why, Massa, you see its worth fifty cents to kill the calf and then there is two dollars for the know how."

PAYING MEN BY THE DAY.

This is another source of loss to both owner and workman, and a constant source of trouble. We therefore advise to give all out by piece work whenever it is possible to do it, and to pay good liberal prices that will encourage good workmen to exert themselves, and every man gets just what he earns and the operator knows every night what his stone has cost through the day. This does away with all complaints about how many hours men shall work per day and many other causes of complaint on both sides.

THE CHEAPEST PLAN OF STRIPPING A QUARRY IS BY HYDRAULIC WASHING.

A book on blasting and mining matters would be incomplete without at least a brief treatise on hydraulic mining or the utilizing of a stream of water under heavy pressure for the removal of gravel or clay deposit. The pressure may be produced by the elevation from which the water decends to the nozzle, this being the general practice in California, and other mountainous regions where streams from great heights are conducted in pipes to the nozzle. To give an idea of the immense power which may be derived from water under pressure, it may be stated that a quantity of water equal to a thousand miners' inches can be discharged under a pressure of say three hundred feet through a six-inch nozzle, with a velocity of one hundred and forty feet per second, and in a volume of one thousand six hundred and fifty pounds during the same period of time. Such a volume, uninterruptedly striking upon a bank of earth or gravel, having, as it does, one-tenth of the velocity of a projected cannon ball, must necessarily do great execution, and produces the caving of an ordinary gravel bank without the necessity of resorting to explosive blasting. The great pressure makes the nozzle difficult to control. To overcome this difficulty, a nozzle in the shape of a goose neck placed upon a universal joint is used and controlled by a hand lever extending several feet to the rear. This device is known as a Hydraulic Giant, and enables one man to direct the stream to do the most effective work. Where the water pressure necessary for this can not be obtained by gravity, pressure pumps are often resorted to with equally profitable results, and a number of these pumps have lately been built by the Hall Steam Pump Company, of Allegheny, both for mining of the precious ores and for the reduction of gravel and clay banks in railroad cutting, and who will no doubt be pleased to furnish to parties interested further details on this subject.

I am strongly of the opinion that many limestone quarries in Lawrence county, Pennsylvania (which have been or are to be abandoned because the stripping is sixteen to twenty feet thick), could be successfully operated by constructing a dam in a small stream near by, and using a properly constructed steam pump to throw the water thus collected against the bank of stripping, and a properly constructed flume or tail race to guide the washings to fill up and level unsightly and useless swamps or ravines below, and at the same time the stream of water could be so regulated in its descent after striking the bank as to cause it to deposit much fine stone from the strippings, which could be used to good advantage for railroad ballast' or on public roads. I may here remark I have known of much trouble being caused in hydraulic washing for want of a very cheap and simply constructed dam to hold the water and cause it to deposit its burden of alluvial matter before escaping into the stream below. This is usually done by piling up mud *only* to form the dam. Thus figure 10 shows the mud that had been deposited by previous washings has

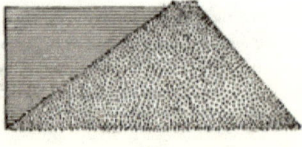

10.   11.

been piled up to form a dam around a deposit field. But as mud is, of course, composed of very fine slippery particles, if the water once overflows at any point it will wash away very quickly and all the water contained in the settling field may be in a few minutes precipitated into the stream below. But if the simple precaution is taken of carrying up a rib of 2x4 inch cheap plank a little in advance of the water, as shown in figure 11, piled up one thickness on top of another, and spiked down with joints broken and then mud banked up on both sides of it, there can be no wash-out. If a simple device of this kind had been used in South Fork dam it might have been prevented, or at least lessened the damage done by the flood.

DYNAMITE AS AN EXPLOSIVE.

Many persons have an idea that because it is *dynamite*, therefore it must do its work wherever put and at an extremely low cost. To all such we would say, there is as much difference in work done by the different grades of dynamite as there is in service done by the different kinds of wood; and any one acquainted with the nature and uses of pine wood would never think of making an ax handle of it; neither would he weatherboard a house with hickory wood. Hence the necessity for studying the work to be done and getting explosives to suit that work. I have had a great variety of experience in blasting almost every known article. While removing the great jam of drift lodged against the stone bridge at Johnston in 1889, where I used about eight and one-half tons of dynamite, and did wonderful blasting among such a variety of substances as to surprise most experienced blasters with the results, sometimes tearing a huge stump to kindling wood, next cutting off a protruding iron bridge girder, next shaking a sunken mass of wrecks of houses all crowded together and interlocked by roots of trees and many miles of telegraph wires. All this gave me great opportunity to study results, and it is my opinion that every blaster should try different strengths and find out what strength does his work best, without regard to whether it is 20, 40 or 60 per cent. Besides, there are no means by which a quarryman can tell whether dynamite really contains 20, 40 or 60 per cent. of nitro-glycerine or not; neither is it of any importance for him to know. All he has to care for is which grade will do his work the cheapest.

During the last ten years there has been a great and increasing

DEMAND FOR HIGH EXPLOSIVES.

During the last ten years there has been a great increase in the consumption of high explosives, caused largely by improvements in there manufacture (making them safer to handle) and increased experience in handling, until now, although a hundred pounds are now used for one pound ten years ago, yet accidents in regular use are so rare, and only occur as the result of gross carelessness, that many users consider nearly all high explosives safer than black powder, and they are now considered indispensable for all wet blasting, such as shaft, incline or well sinking, or taking up bottoms in coal mines or submarine or salamander blasting, or any use where great shattering power is wanted, because high explosives make eight or more distinct cracks, radiating from the hole like spokes of a wheel (see fig. 12), while black powder generally makes but three cracks (see fig. 13).

FIG. 12.   FIG. 13.

When dynamite is exploded in a hole it makes eight distinct cracks or fractures which radiate from the hole same as the spokes of a wagon wheel does from its center, as shown in fig. 12, while black blasting powder makes only three distinct cracks as shown in fig. 13, and while the fragments of rock, after being blasted with black powder, are still solid and can be used for dimension stone. The large fragments, after being blasted with dynamite, are unfit for dimension stone, having received such a shock as to produce incipient cracks, which may cause the stone to break in pieces after much labor has been spent in dressing it. Quarrymen should therefore bear in mind that dynamite should never be used on dimension stone that is wanted solid for dressing, and also that dynamite is the proper explosive to use in quarries where the stone is wanted broken small, as for railroad ballast, concrete foundation, common road metal, etc., etc.

DYNAMITE MAY BE USED LAID ON SURFACE OF ROCK.

Dynamite can be most economically used when placed in a hole and tamped solid, but in exceptional cases it can be used to good advantage when laid upon the surface without drilling holes, as for instance where a rock has fallen down on a railroad and obstructs traffic—where time is of first importance and the cost secondary—a rock can be broken quickly by laying four or five times as much dynamite on top of it as would be required to blast it if put in a drilled hole in it; when thus used all charge should be covered with five or six inches of mud; and for breaking cast iron a very small charge will be sufficient to break it at one-fourth the cost it would take to drill a hole in it, and to cut up old heavy scrap iron, like old steam boilers, it can be done very cheaply by simply spreading out a strip of paper four inches wide and as long as the cut is desired to be in the iron, then on this paper lay a thin strip of dynamite about one inch wide and one eithth of an inch thick, which has been crumbled out of a cartridge, then fold the edges together so as to bring the dynamite together in a continuous long roll or ribbon, with the edges of the paper folded under each other (as a druggist folds up a prescription) so as to hold the dynamite in a continuous ribbon or roll and to prevent the wind from blowing it off; then lay it over the place it is desired to cut, then place a fuse two feet long (with explosive cap attached) about the center of the ribbon; this prepared, then place a teaspoonful of dynamite over the cap, then cover all two or three inches deep with mud, then fire the fuse and get at least two hundred feet away, and as soon as the fire burns up the fuse till it reaches the cap it will explode and the cut will be made. A little experience will soon enable any person of ordinary intelligence to reduce or increase the charge so as to do it most economically.

DYNAMITE IS INVALUABLE FOR BLASTING STUMPS.

For blasting stumps and rocks in fields, many thousands of acres in Pennsylvania have within the last ten years been cleared of stumps and boulders, to the great satisfaction of their owners, as far cheaper than any stump machines, and at the same instant that dynamite knocks the stump out of the ground it splits and breaks it in pieces for easy handling, drying and burning; it makes but a small hole in the ground, and many farmers have told me that the first crop from the ground formerly occupied by a stump has more than paid all expenses.

DYNAMITE IS NOW MADE PERFECTLY SAFE

To transport or handle, and nothing but pure carelessness and utter neglect of simple instructions can cause any damage while using it. For instructions about thawing dynamite, see page 12.

STUMP BLASTING MADE EASY AND SAFE.

FIG. 14.

PREPARING STUMP FOR BLASTING BY USING HIGH EXPLOSIVES.

DIRECTIONS FOR BLASTING STUMPS.

1st.   Cut off about three feet of fuse (cutting it straight across), then shake all sawdust out of a cap and push the fuse into the cap gently as far as it will go, then with nippers or dull knife close the cap down tight on the fuse.

2nd.   Open the one end of a dynamite cartridge and with a pointed piece of wood about the size of a lead pencil, make a hole in the powder large enough to receive the cap, push the capped end of the fuse down into the powder, close the paper around the fuse and tie the fuse to cartridge, so that it cannot pull out.   See page 27.

3rd.   With clay auger, or pointed crowbar, make a hole in the earth down under center of stump, place the cartridge at extreme depth of hole, leaving the fuse extending out above ground, then ram the earth in tight around cartridge and up level with top of ground, fire the fuse and get away two hundred and fifty feet and watch results.   Inside of two minutes the stump will be thrown out, torn to fragments, which can easily be piled up, dried and burned out of the way.

One such cartridge is sufficient for any stump one foot in diameter ; larger stumps will require more. A little experience will soon show how much is needed, as it takes more to some kinds of wood and localities than others. Any number of cartridges can be fired by one fuse and cap, provided they all touch each other. Place the cartridges from twelve to twenty inches below bottom of stump in the ground.

FIG. 15.

Figure 15 represents a piece of one inch pipe, five feet long, closed at one end and pointed at the other, which makes a most excellent tool for probing among roots, feeling for an opening for the auger (fig. 16), which I have found by experience to be the best way for making the hole for the blast.

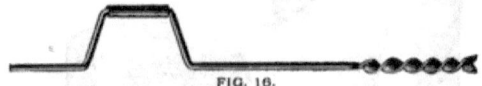

FIG. 16.

Figure 16 shows an auger for making holes under stump, which is the best plan I know of for that purpose.

TO BLAST A LARGE PINE OR HOLLOW STUMP.

Place a strong chain around stump, key it tight and then drive in tightening wedges until the chain is tight around the stump, and as near the top of stump as strength of wood will permit. Then place an extra heavy charge, as shown in figure 14. In all cases it will be found cheaper to have the charge a little too strong and throw the stump out at once, all torn to fragments, than to have to use a team of horses a half hour to drag out roots.

TO BLAST FRESH CUT OAK OR HICKORY STUMPS HAVING A LARGE TAPE ROOT.

With auger (fig. 16), make a hole down along side of tap root, nearly three feet deep; enlarge the hole at bottom by holding auger to one side, then having the dynamite thawed (if in cold weather), cut the cartridges into three pieces and slit the paper wrapper from end to end. Place cap and fuse inserted in one piece of cartridge at bottom of hole, then place one-third of dynamite cartridge in hole and bruise it solid with a wooden rammer, then another and ram it solid, and so on until you have what you think enough; then tamp up solid to surface of ground and fire, and get away at least two hundred feet, or if there is water in the hole let it fill up and it will take the place of tamping and cutting the cartridges.

HOW TO USE DYNAMITE.

Dynamite is generally put up in cartridges of strong manilla paper, each cartridge eight inches long by one and one-quarter inches in diameter, but when wanted in smaller holes the paper cartridges can easily be opened or cut, and the dynamite can then be crumbled into the smallest holes, but if several hundred pounds is to be used in this way, it is better to send an order to the factory and have the cartridges made the exact size. It is very important to have the charge packed solid in the bottom of the

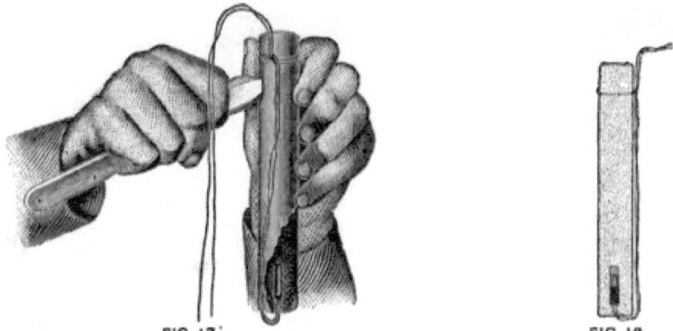

FIG. 17.   FIG. 18.

FIG. 17 SHOWS HOW TO FASTEN AN ELECTRIC EXPLODER TO A DYNAMITE CARTRIDGE.

hole, and in order to do this I recommend the blaster, after he has inserted the fuse, as described on page 26, to then place the nail of his right thumb on the side of a sharp-pointed pocket knife, about one-quarter of an inch from its point, and then (fig. 17) insert the point of the knife into the cartridge near one end and then draw the knife to the other end of the cartridge, and thus slit the wrapper from end to end, then lower the cartridge to the bottom of the hole and with square cut the end of tamping stick, push the cartridge down solid to the bottom of the hole, if the hole is to be fired by electricity, and insert the electric exploder in the first cartridge, as described on page 23. But if fuse is used, insert the exploding cap in last cartridge of the charge of dynamite, but in firing black powder it should always be fired at bottom of charge. To fire by electricity is by far the best plan (but whether ordinary fuse or electricity is to be used, each mode will be described under its own title). **First open one end of dynamite cartridge** and with a pencil-like stick (made of hard wood) make a hole in the end of cartridge sufficient to permit the whole cap to be embedded in the dynamite, then close the end of cartridge, and if electric wires are used, bend the wires back and make a half-hitch around the other end, as shown in fig. 17. But if fuse is used in a pop hole, not over one foot deep, the capped fuse can be pushed into the dynamite sufficient to lower it down to the bottom of the hole without pulling out, as shown on page 25, or the capped end of fuse may be placed at the bottom of the hole and the dynamite crumbled into the hole and rammed tight with a wooden rammer and tamped tight in the usual way with black powder.

HOW TO FIRE BLASTS IN COAL, ROCK, STUMPS, ETC.

There are three different plans of firing blasts in coal, rocks stumps, etc. First, squibs. Second, fuse. Third, electricity.

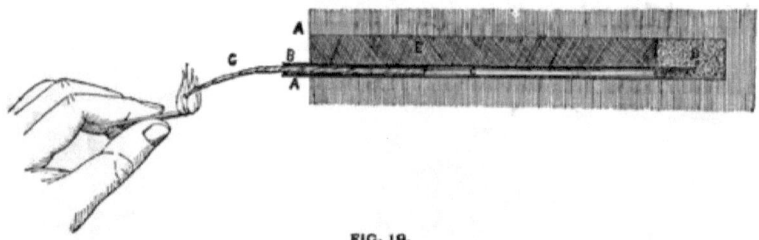

FIG. 19.

First, I will describe squibs, which are artificial straws, about four inches long, made of paper and filled with rifle powder and closed at one end by a small plug of resin soap, and the other end has about three inches of a slow fuse attached, made of paper impregnated with a solution of saltpeter and dried, then cut into suitable strips and pasted around the other end of squib and then dipped in melted brimstone. Squibs are used almost exclusively by coal and fire-clay miners in places where fuse would make too much smoke. Fig. 19 represents a blast ready to fire in coal. AA represents the coal through which hole B has been bored horizontally one and one-half inches in diameter and four feet deep ; D is the powder which has been put there by means of a powder spoon (fig. 22½, p. 26), and C is a blasting barrel which is simply a plain piece of quarter-inch gas pipe, four feet long, which is placed in the hole after the powder (D) has been put in, and before the tamping (E) has been put in; B is the tamping, which is generally the borings out of the hole rammed tight in again; G is the squib placed loosely in the end of blasting barrel (C), and is ready to fire when the fuse end is fired. It will burn one minute before reaching the powder. This gives all persons time to get away before it burns to the powder, when the recoil of the burning powder will project the burning squib to farther end of blasting barrel at D and fire the charge. It will easily be seen that there is a very serious objection to this mode of firing, because much of the force of the explosion must be lost by escaping through the pipe (C) in an ignited but only partially consumed condition, thus wasting powder and producing far more smoke than would result if the hole had been fired by electricity.

FIRING BY FUSE

Is so well known to most quarrymen that any description would appear useless, but as there may be new beginners looking for information, I submit the following : Fuse sixty years ago was made by hand by the quarrymen in a great variety of ways, but it is now made exclusively by factories supplied with suitable machinery to make it in continuous rolls of fifty feet each, and is now made of four descriptions,

GUTTA-PERCHA.

DOUBLE TAPE.

SINGLE TAPE.

COTTON.

HEMP.

FIG. 20.

namely : **Cotton, hemp, single tape and double tape** fuse (and formerly guttapercha, which is now seldom used). All of these makes of fuse consist of a continuous cotton thread in center, which has been run through a solution of saltpeter and dried; this greatly increases its fire-carrying power, which is placed in the center of a column of fine rifle powder of about one-tenth of an inch in diameter, and this column of powder is wrapped closely and spirally around with eight very coarse soft

FIG. 21.   FIG. 22.

cotton threads, and this is again wrapped in opposite direction with five doubled and twisted cotton threads, and after this it is coated with a tarry substance to make it partly water proof and to hold the powder in, the whole being, when finished, about three-twelfths of an inch in diameter; this is used only in dry holes and is called cotton fuse, and is too small to fill the exploding caps in dynamite blasting. Hemp fuse is made in the same way as cotton fuse, except that eleven threads of hemp take the place of the eight coarser cotton threads and the fuse, when finished, is a little smaller than cotton fuse. Single and double tape fuse has the same saltpeter-impregnated cotton thread in its center, surrounded by a tube of fine powder one-tenth of an inch in diameter, and is then surrounded by sixteen fine cotton threads wound spirally around it, and again by five doubled and twisted cotton threads wound in opposite direction to that of the last five threads, and this is again wound in opposite direction by ten threads of hemp wound spirally around it, and this again is covered by a winding of cotton tape five-eighths of an inch broad and laid on so that it comes almost edge to edge. But in double tape fuse, this tape is made to describe its spiral motion so slow that it overlaps so much that there is always double thickness of tape around the fuse; then on top of all this, is a coating of pitch to make it water-proof. When you bear in mind that the cord or tube of powder in all these fuses burns at the rate of 3 feet in one minute, one can easily see it is quite possible that among all this mass of burnable material there may still remain some fire, which I have demonstrated by actual experiment, wil continue to burn ot the rate of four inches per hour and still carry fire sufficient to fire a charge of powder while squibbing a hole.

HOW TO FASTEN FUSE TO CARTRIDGE

For deep hole blasting is shown in fig. 23, where it can easily be seen that the twine has been passed around the cartridge about one or two inches from one end and tied to the fuse so that the two cannot be separated in the hole. Of course the fuse has its cap on and the paper will now be slit from end to end, all as directed on page 23, and lowered down to bottom of hole and bruised solid as directed. Fig. 18 shows

HOW TO FASTEN AN ELECTRICAL EXPLODER TO A DYNAMITE CARTRIDGE

Without the use of twine, by simply making a hole in the end of the cartridge with a pencil-like stick of hard wood, pointed at the end, and one-eighth of an inch larger than the cap. Let it be of hard wood and so sharp pointed that you can easily punch it through the paper without taking time to open the folds of paper on the end of cartridge ; withdraw the stick and insert the cap of electric exploder in the hole so as to have the cap buried at least half an inch deep over all in the dynamite, then turn the wires up alongside the cartridge and take a half-hitch around end of cartridge, as shown in fig. 18 and it is then ready to be lowered into the hole.

FIG. 22½.

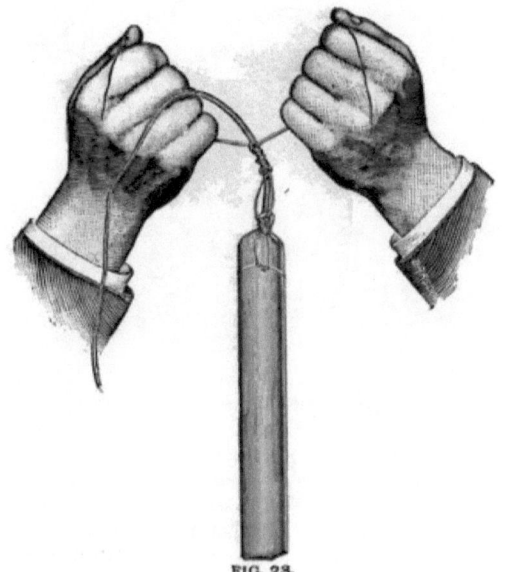

FIG. 23.

Figure 23 shows how to fasten the fuse to a stick of dynamite, so that it will not pull out, and if the dynamite should not go off right it can be withdrawn and put down again.

BLASTING BY ELECTRICITY.

Economical Value of Blasting by Electricity.

Those who have most carefully studied the matter are most earnest in praise of the method and its economical value. A very little thought will make apparent the greater effect which can be produced by firing *simultaneously* a number of contiguous blasts, instead of firing them singly, while a little experience will teach that even in firing single blasts by this apparatus, much can be gained. One advantage gained in firing single holes by electricity is, that in case of misfire (which can rarely happen by this method) no time is lost in waiting, as in the case of firing by safety-fuse, for there is no hanging fire with electricity. Another advantage is, that the explosion of the electrical fuse at the bottom of the charge throws the fire through the whole body of the powder, igniting it all at once, and by *detonation*, giving the same charge far better explosive effect, as has been fully demonstrated by experiment.

Another advantage is, the explosion of an electric fuse at the bottom of charge produces the first rupture in the rock at bottom of the hole, leaving the tamping perfectly tight, and confining all the explosive gas until the powder is completely comsumed and its force expended on loosening rock.

FIG. 23½.
MAGNETO MACHINE.
5x8 Inches and 14 Inches High.

When firing is done with a common fuse, the first point of rupture in a rock is often near the top of charge, which often leaves a large part of the hole unruptured, the drilling of which, and the powder it contained, were a total loss, for if it had exploded it would have burst the hole.

Electricity is much safer than fuse for blasting among houses or along railroad tracks, because it can be done in one moment when everything is safe, and as it fires at bottom of hole it will not throw so much material around on houses, and there cannot be any hang-fire shots to endanger the public.

COST OF ELECTRIC APPARATUS.

The Magneto Electric Machine is now the best in market, and costs only $25.00 and will fire one or forty holes on separate charges (fig. 23½).

LEADING WIRES

Cost different prices, according to quality, from one cent per foot up. But after many years' experience, I recommend the following outfit: One battery, one reel and two hundred and fifty feet of double braided cable, having great carrying power and very durable, and very easily taken up out of the way of workmen, carts, etc. The whole now costs $40.00.

BLASTING REELS.

The want of some convenient contrivance for handling and shifting wire from blast to blast has long been felt by those using electric batteries for blasting, and I therefore take great pleasure in calling attention to an improved reel, as shown in fig. 24. It consists of a strong hard-wood frame with a reel

FIG. 24.

CRESCENT BLASTING REEL.

After much experimenting I have found the reel shown above to be the best I have ever seen used.

inclosed. It is but the work of a minute to run out as much leading wire as is needed to reach a place of safety for reel, battery and operator, and connect to battery and fire, and as quickly wind up again. There need be no loss of time or wire and it is my opinion that more than cost of one reel will be saved to the blaster in saving of time every month of its use

DIRECTIONS FOR USING CRESCENT BLASTING REEL.

First, strip all insulation off three or four inches from the end of each of the cable wires to C, fig. 25; then pass one of the wires from the inside of the reel through one of the holes (E) near center of reel (see figs. 24 and 25); then pass it back again as in fig. 25. A, fig. 25, represents sheet iron end of reel with

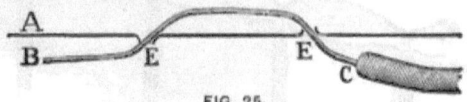

FIG. 25.

two holes (E E) through it near its center, and B, the end of wire with insulation taken off to C, and the naked wire (B) has been passed from the inside of end of reel through hole E, and back again through hole E to inside end of reel; then do the same with the end of other wire from double cable, passing it through holes EE in the other end of spool, and then by means of the crank wind all the cable on reel, as shown in fig. 24, and it is all ready for use.

DIRECTIONS FOR CHARGING HOLES TO FIRE WITH ELECTRICITY.

First, select an electric fuse at least four or more inches longer than depth of hole, then make it fast to a dynamite cartridge, as shown in fig. 18, and slit the paper as directed (page 23), and lower the cartridge by the wires of the exploder to *bottom* of hole; then hold the wires in one hand and with smooth, square-cut end of tamping stick bruise the cartridge down so as to fill the hole tight; then add as much explosive as will do the work desired, and tamp same as for fuse (but all the time you should hold both exploder wires in one hand against one side of hole to prevent them being buckled or cut).

DIRECTIONS FOR BLASTING BY ELECTRICITY.

1st. Drill as many holes as you want to fire at once; each hole should not be more than its own depth from the next hole; thus, if the holes are six feet deep, they should only be six feet apart, and six feet from the face of rock.

2d. Cover bottom of hole one inch with powder, select a fuse long enough to reach bottom of hole, and leave three inches or more out of hole; place the exploding cap *at bottom of hole*, and then put in the usual amount of either black powder or high explosive, then tamp in the usual way, taking care not to rub or buckle the wires in tamping.

ELECTRIC FUSE. FIG. 26.

To connect wires, prepare the ends to be connected by having two inches of clean, bright wire, without any insulation on them, then take one wire in each hand and cross them at about one and a quarter inches from the ends, then twist them eight or nine times (see fig. 28).

When all the holes to be fired at one time are tamped, separate the ends of the two wires in each hole, joining one wire of the first hole by means of connecting wire with one of the second, the other, or free wire of the second with one of the third, so proceeding to the end or last hole, thus:

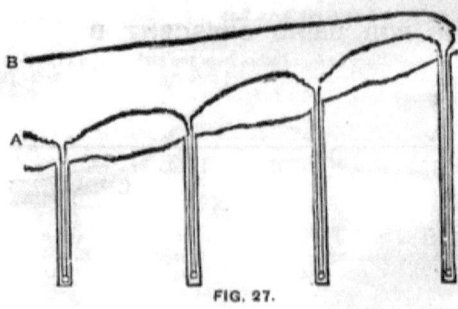

FIG. 27.

30

HOW TO CONNECT THE WIRES FOR BLASTING.

Presuming that you have done as directed above and got your holes (any number from one to thirty-five) all tamped and ready with wires all projecting above the rock four inches or more, you now begin at one end of your line of holes and part the two wires in first hole, and splice a piece of connecting wire to one wire of first hole, fig. 27, A, leaving one wire of first hole disconnected for the present ; you connect one wire of first hole to one wire of second hole ; and here I should remark that all connections must be made with naked wire to naked wire ; no insulating material should be on either wire of a splice, but the rule naked wire to naked wire, must be observed in every splice, thus :

FIG. 28.

Having spliced your connecting wire to one wire of first hole as above, you cut off enough connecting wire to reach the next hole, and clean off the insulation from its end for two or three inches and splice it to one wire of second hole, and thus proceed until all holes are thus connected. You will then have one wire of first hole and one wire of last hole still unconnected ; then bring wires D and C from the reel and splice D to one free wire of first hole and wire C to free wire of last hole, and then carry reel to place of safety (p.33), permitting the cable to run off as you go ; connect K from the battery to holes A in end of reel, and wire L from battery to holes B in reel ; then make sure that every person is out of danger, then shout Fire ! and take position, as shown in fig. 30, and draw up wooden handle and place both hands on it with each foot alongside of battery and body over it as shown in the cut ; then start to push the handle down gently at first for the first inch and rapidly increase the downward motion without stopping or slackening speed, but come down to end of stroke with a sharp thud, and as soon as the lower end of brass rack bar strikes a spring at the bottom every hole will fire, but in case it don't fire, remember the first thing to do is to disconnect one wire from battery (every time you try the battery) and then go over the whole circuit of wires and see that every splice has been properly made ; and the best way to do this is to begin at the battery. First, see if any defect in the wires exist near the battery and as soon as you find anything wrong, correct it, then come back to battery and connect the wire you loosened from the battery and fire, as already described. Remember the importance of seeing that all connections on an electric circuit must be made with naked metal to naked metal. Then if the charge will still not fire, begin at the battery and run your hand along all wires around the whole circuit and back to the battery again, feeling the wire as you go along to find whether or not it has been struck by a falling stone from the last blast and the wire cut, but still held in position by the insulating wrapping ; and as you pass along examine every splice. As you pass along over the wires between holes, pull them gently tight to find whether or not some person has caught a wire on his foot just as he was leaving before the blast and broken the wire. But if the blast will still not fire, first thing to be done after firing, is to disconnect one wire from the battery. Then disconnect one hole at one end of the blast (which we will call No. 1 hole, for ease of description) and leave it out and connect the wire from battery to free wire of hole No. 2 ; then operate the battery again as already directed, and if it still will not fire, then cut out hole No. 2 and connect to hole No. 3, and try the battery on it, and so on until you get the remaining holes to fire; then leaving out the last hole, connect all the remaining holes together and fire them, and then attach each of the cable wires to one of each of

31

FIG. 29.

the wires in the hole and operate the battery as already described, and if that hole will not fire then drill a hole, say ten inches from the unexplodable hole, and charge and fire in the usual way, and then search carefully for the cause of misfire by examining the wire before any one has a chance to disturb it, and most likely you will find the person who loaded the hole did not hold the wires in his hand, but permitted them to be doubled up in tamping the hole. If no apparent cause can be found for misfire, and you can get the exploder out that should, but did not explode, then attach one of its two wires to one wire each of the cable and operate the battery, and if it fires then, there must have been something wrong in tamping, and if it will not fire, then return it to the maker as a bad and condemned exploder.

THE IMPORTANCE OF THIS SEARCHING INQUIRY

Although long, will appear when you remember that I am writing for the information of new beginners in using the battery, and wish to give them all the information I can, for I have repeatedly had complaints of great trouble where the information given above would have saved great trouble and expense.

FIG. 30.

ANOTHER CAUSE OF MISFIRE IN ELECTRIC BLASTING

Is caused by the reckless handling of electric fuses or exploders before they are used. I have repeatedly found a bunch of them all tangled up like a handful of hay, and men tugging and pulling at them as if they were so much brush from the woods. In doing this, I have no doubt, they often disturb the insulation of the wires, and thoughtlessly cause great trouble and loss, as above.

THE GREAT PERFECTION OF MACHINERY FOR MAKING ELECTRIC EXPLODERS,

And the great care now taken to test every separate exploder before it leaves the factory, renders it impossible that a bad exploder should ever get upon a quarry. But if one ever is found that will not explode, the maker takes it as a great favor if it is returned to him, so that he may examine it and do something to prevent another one getting out that will not explode.

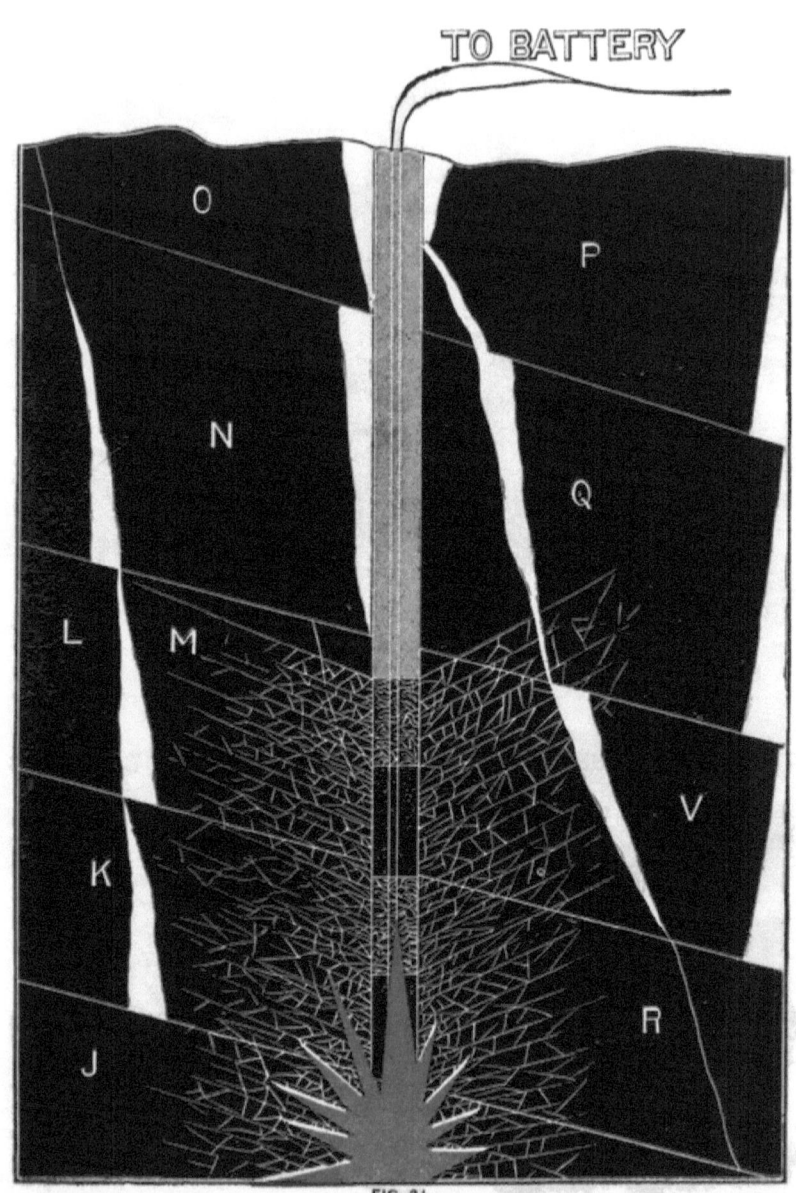

TO BATTERY

FIG. 31
OLD PLAN OF HEAVY BLASTS.

34

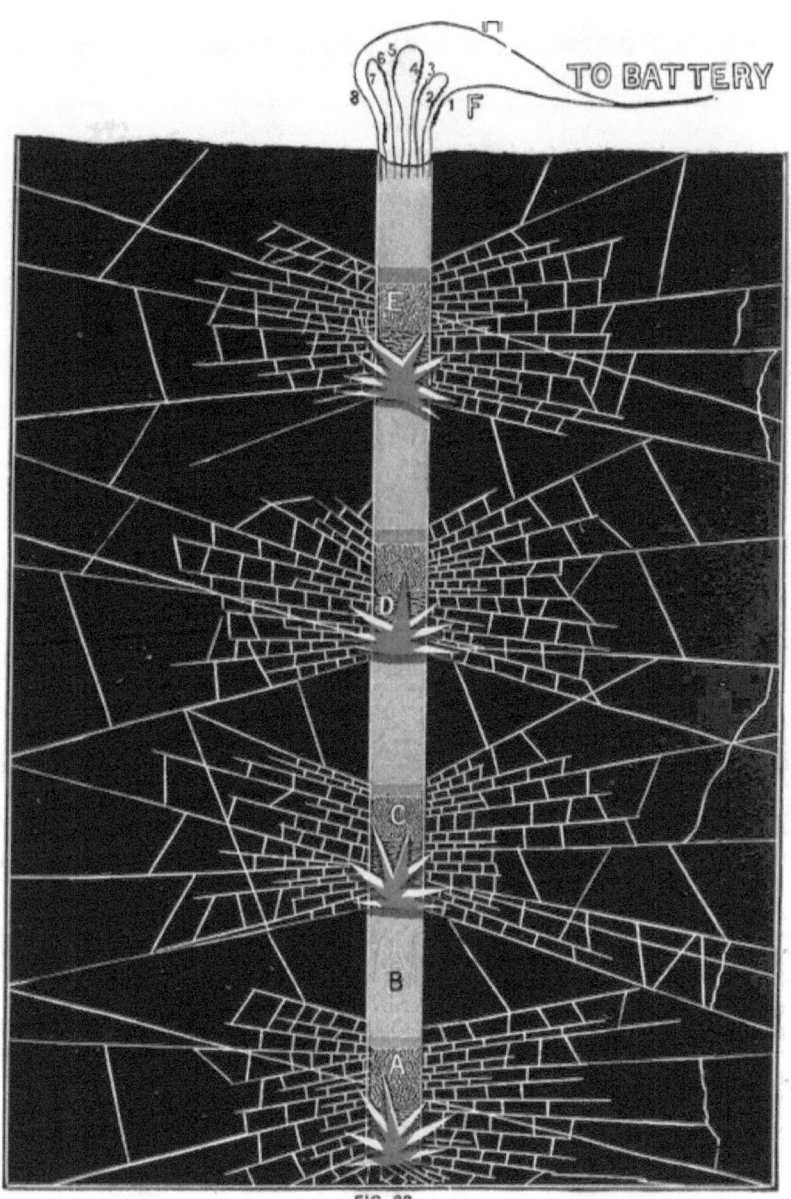

TO BATTERY

FIG. 32.
NEW PLAN OF HEAVY BLASTS.
35

CONTRAST BETWEEN OLD AND NEW PLAN OF BLASTING.

Figures 31 and 32, pages 34-35, show the contrast between the old and new plan for making a blast in a hole, say twelve or sixteen feet deep in limestone, shale, etc., where it is desirable to break up the rock into moderate sized pieces and yet not waste rock; and here I would remark that taking advantage of the facts already described, relating to the shivering or fracturing properties of dynamite and the expansive or propelling power of black blasting powder, I have shown in both figures how to combine both together by first putting in, say, twelve inches of dynamite at the bottom of the hole (fig. 31), and placing the electric exploder at bottom of charge, so as to produce the first rupture at the very bottom of the hole; then, say twelve inches of black powder, then twelve inches of dynamite (as shown in fig. 31), and so on, until eight feet of the hole is thus packed tight with alternate layers of dynamite and black blasting powder, and eight feet of the hole is tamped solid. Thus far applies to both holes alike. But in fig. 32 it will be seen I cut off the charge, say every two feet, by putting in two feet of combined explosive at A, then two feet of tamping B, then two feet of explosive C, then alternate layers, two feet of each, to top of hole, with an electric exploder in every charge of explosive, the wires of which are to be connected together at top as if each exploder had been placed in a separate hole. Thus, wires 1 and 2, at top of hole belong to exploder in charge A, and wires 3 and 4 to exploder in charge C, and wires 5 and 6 to exploder in charge D, and wires 7 and 8 to exploder in charge E. Now, if we leave wire No. 1 free to be connected by cable wire F to the battery, and connect 2 with 3, and 4 with 5, and 6 with 7, then 8 will be left free to be connected with wire H to battery, or to wire 1 in an adjoining hole, and any number of contiguous holes may thus be connected until their combined number of exploders amount to thirty-five for a No. 3 push battery, or sixty for a No. 2 battery.

THE RESULT OF THE TWO MODES OF FIRING.

The results of the two modes of firing is very clearly shown in the two cuts, figs. 31 and 32. In fig. 31 all the explosives having been confined in one continuous hole with all the mass of inert stone above it has so burned and shattered the stone near it, that I have seen blasts made where several cubic yards of good limestone have been so shattered, burned and blown away as to be a total loss to the quarrymen, and at the same time many blocks like J, K, L, M, N, O, P, Q, V, R, left unbroken. I have seen such blocks of all sizes up to sixty cubic yards. These blocks must then be drilled and reblasted at great loss of time and expense for explosives. But in fig. 32, where the shattering force of dynamite and propelling power of black powder has been more generally distributed through the mass of rock by simply placing two feet of tamping between every two feet of explosive, here the rock is far more equally broken up, less is burned, shattered, blown away and lost, and there is less need for pop hole blasting, and the result will be a great gain to the quarryman or contractor by not wasting rock near blast as above, and breaking up the rock more generally, saving stone, time and expense for pop hole drilling and explosives.

BLASTING BY ELECTRICITY COMPARED WITH FUSE BLASTING.

And by means of it fire a long range of holes at once and have every hole explode at one and the same instant, and thus remove double the quantity of rock than can ever be done by fuse blasting with the same number of feet of hand-drilled holes, as will be clearly understood by referring to following cuts (see fig. 33) where the line A B C represents the face of the rock in a quarry or side cut before blasting, and if

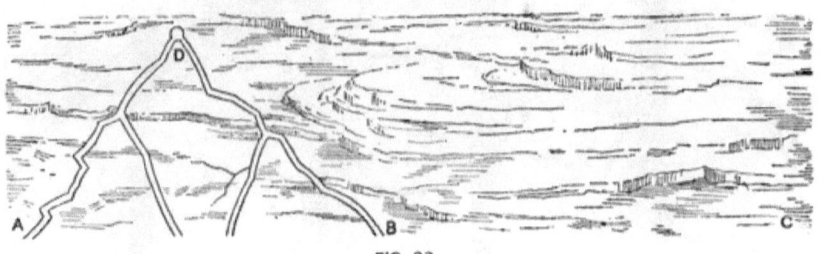

FIG. 33.

hole D is drilled by hand, say ten feet deep and ten feet back from face of rock A B and then fired by fuse in the ordinary way after squibbing it twice, perhaps enough powder can be put into it to break out the rock to lines represented by A D B. Then hole G (fig. 34) can be drilled, for in seamy rock with fuse blasting, G cannot be drilled until D is blasted, for fear that D may break into G. But in electric blasting this can make no difference, because the explosion takes place in every hole on the circuit at the

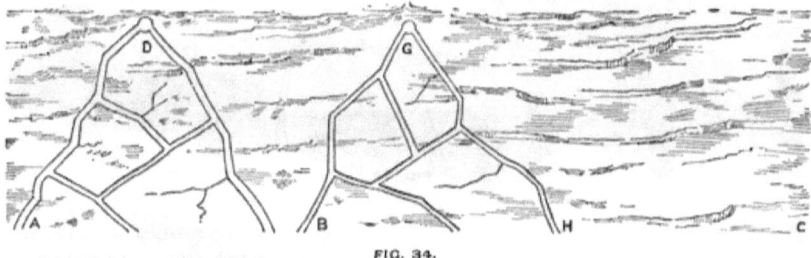

FIG. 34.

same instant and the force from any one hole meets the force from its neighboring hole like two bars of iron butting against each other, and their united force is exerted on the rock at the same instant. When G is fired it breaks out to B G H, and when I is fired it breaks out to H J K (fig. 35), leaving the points A B H unexploded; but if these same holes had been fired by electricity, the points A B H would be gone as shown in fig. 35, and thus nearly double the amount of rock can be loosened by firing the same holes and same amount of explosives if a range of contiguous holes are fired simultaneously by electricity instead of being fired in the old way with fuse, only a single hole at a time. It is almost impossible to properly estimate the great advantage in firing large blasts at once and thus loosening up a large

37

amount of stone at once, saving loss of time of a whole gang from five to fifteen minutes every hole that is blasted and besides the men having plenty of stone to load from can work to far better advantage than if the holes are blasted only one at a time. The new face of the rock would have been along the dotted line A D B G H J K (fig. 35), but the points A B H having been all blasted off they make nearly

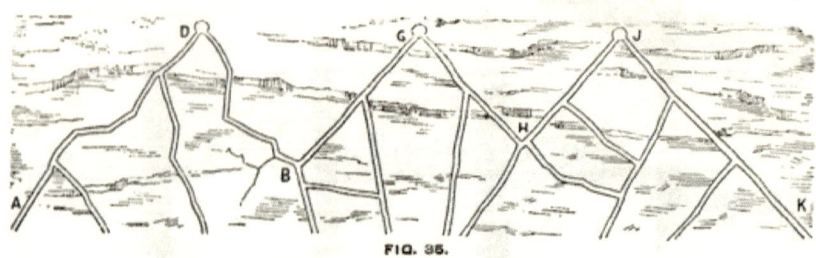

FIG. 35.

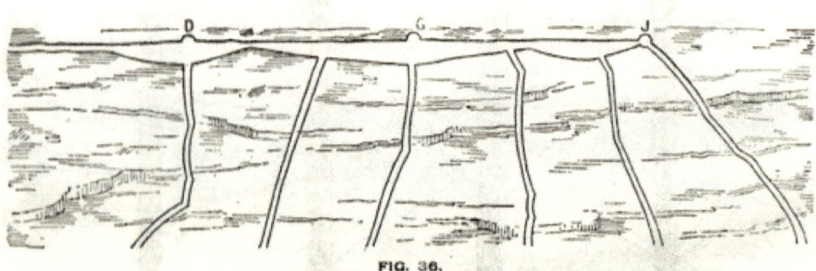

FIG. 36.

double the amount of rock removed by electric blasting than if same holes had been fired by fuse, because while it may be ten feet to top of rock and ten feet to front of rock, yet it is only ten to twelve feet from hole to hole; then as the explosion takes place in every hole on the circuit at the same instant, the force of explosion has only to force six to seven feet in direction of the next hole (fig. 36), until it is met by the force of explosion in adjoining holes, thus making first rupture in the rock from hole to hole, and the expansion of the blast is spent in enlarging this rupture, and yet another advantage of electric blasting over fust blasting is **producing the first point of rupture** at the very bottom of every hole on the circuit at the same instant, as is shown in fig. 37.

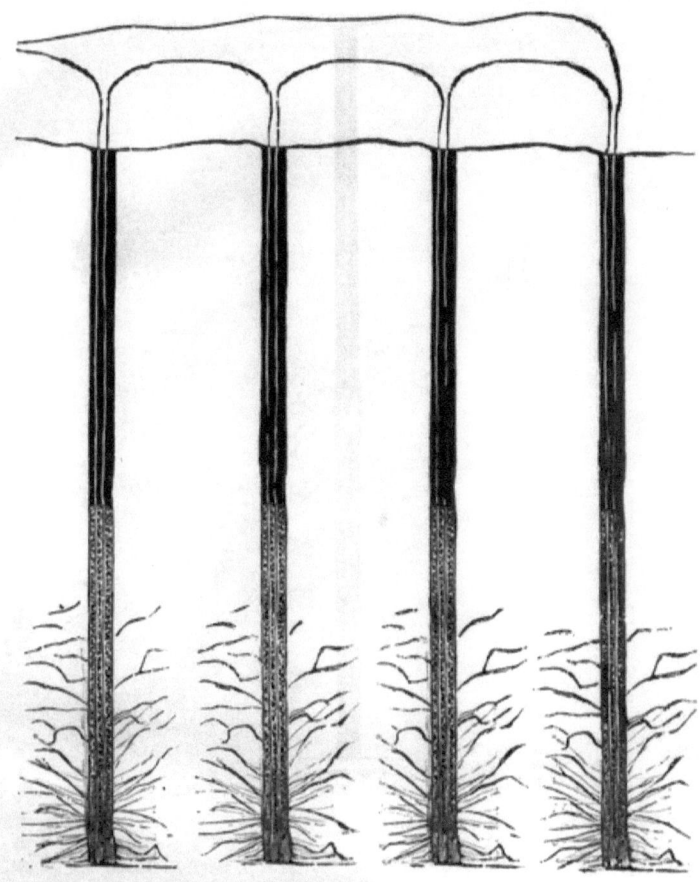

FIG. 37.

These four cuts on this page represent a line of holes, which may be any number up to forty holes, all at one shooting with a No. 3 Magneto Electric Battery, and sixty holes with a No. 4 Electric Battery and at any depth, and of different depths, yet they can all be easily fired. A smart movement of such a simple machine as the battery (which will be explained further along) and the above holes can all be fired at same instant, thus securing far better results than could be done by forty shots fired by fuse, besides being much safer, for if by some carelessness in loading the holes one hole or more fails to explode, there is no **hang fire or no danger** in going up to it or working around it, as there is no fire in the hole and nothing but electricity or a hard blow on the cap can fire it.

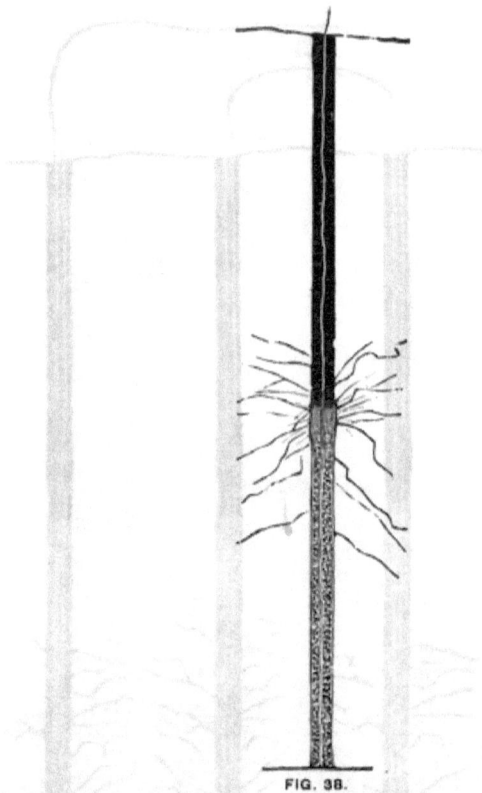

FIG. 38.

Supposing the two holes A 1 and B 1, fig. 39, to be any given depth, say ten feet deep each, and each charged half full with powder, A 1 having an electric fuse and B 1 a common fuse, as soon as the fire reaches the powder chamber it ignites the powder in B 1, say four feet from bottom, and the first point of rupture is, say four feet from bottom of hole; hence we often find from one to two feet of hole unexploded (see B 2, fig. 39). When this is the case all the labor spent drilling that stump of hole was lost, and all the powder it contained was lost. It may often be seen going off in a second flash after the first fire, and a great part of the stone blasted is scattered around and lost.

But firing with electricity never leaves any stump of hole, because the first point of rupture is always at bottom of hole, and the powder being confined under the whole load of rock (see A 2, fig. 39), the full force is expended in enlarging the first rupture, and no stump of hole is ever found after electric blasting, the rock being always loosened to bottom of hole.

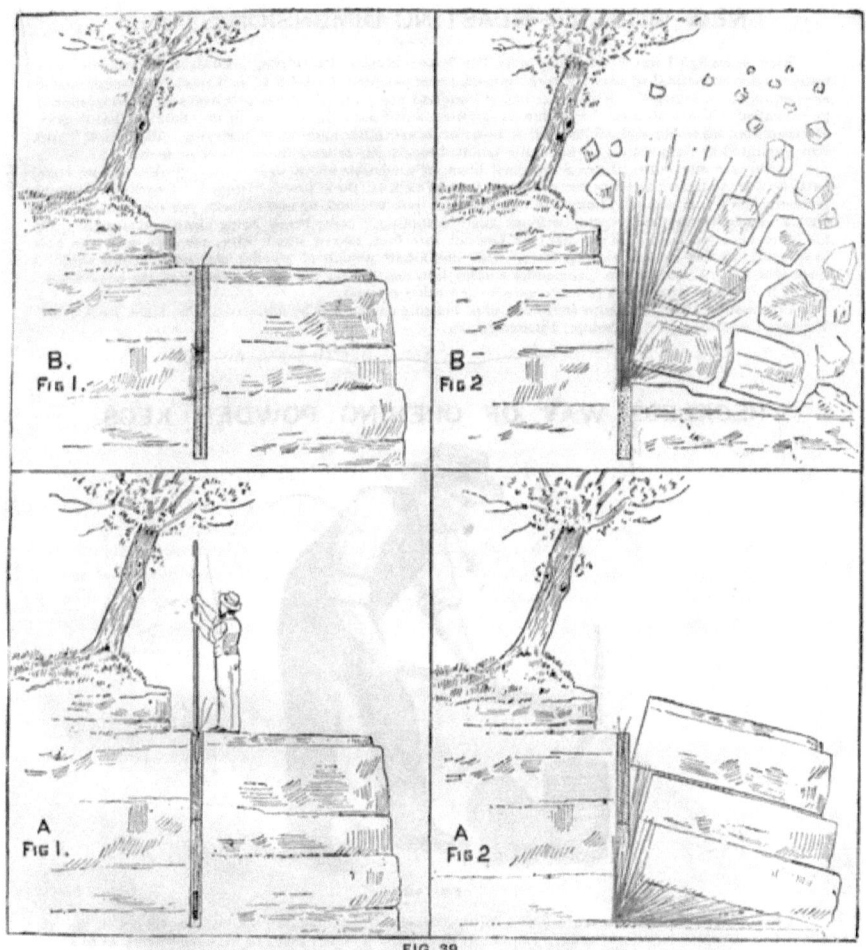

FIG. 39.

The advantage of electric blasting over fuse blasting is still farther shown above, where fig. A 1 shows a hole just loaded and tamped to fire by electricity (fig. A 1), the same hole fired and as the first point of rupture took place at the bottom of the hole, the rocks are merely moved out of their bed, neither broken up too much or scattered around. B 1 shows a hole loaded and tamped to fire by fuse and B 2 shows the first point of rupture was made in the rock some two feet from bottom of hole, and the rock is all smashed into fragments, much of it being scattered around and much of it broken so as to be lost or worthless.

NEW MODE OF BLASTING DIMENSION STONE.

Some years ago I was present at a quarry in Beaver county, Pa., during a series of experiments illus-
trative of a new method of blasting rock, invented and patented by John L. J. Knox. The experiments
were eminently successful. Since that time I have had many opportunities of witnessing the operation of
the so-called "Knox Method" in different quarries, and I am still, as then, of the opinion that in point
of expedition, economy and efficiency it is superior to any other method of quarrying "dimension" rock
ever presented to the public. If judicially handled no danger arises either to rock or men.

In the case referred to, I saw a detached block of sandstone about 14x10x2 feet broken in four equal
parts by a single hole placed in the center, and then each of these pieces "ripped" lengthwise through
the center by a single (inch) hole with one ounce of powder, making eight blocks 7x2 feet 6 inches by 2
feet, all ready for the stone cutter without any "scabbling," every break being clean and straight. On
the same occasion I saw a block 44x24x8 feet cut into four, two of which were 12x12x8, with one hole
(1½ inch), all the breaks being perfect. One and a half pounds of powder was used in this shot. I
learn that the "Knox System" is rapidly coming into use all over the country, many of the largest companies in the country using it in preference to any other method.

Information about this improved method of blasting can be had by addressing The Knox Rock Blast-
Blasting Co., No. 95 Fifth avenue, Pittsburgh, Pa.

RECKLESS WAY OF OPENING POWDER KEGS.

FIG. 40.

I stop the press to call attention to a reckless mode of opening powder kegs, as shown in fig. 40,
which I have reason to believe is very commonly done, and how any one can be so foolhardy as to strike
a steel pick through the sheet steel end of a powder keg, when that keg is full of powder up to about one
half inch of the keg head, is more than I can understand, for the smallest spark from the end of a pick
would be sufficient to fire the whole keg, and that would fire all around, and yet I assure the reader that
I have seen this done by a quarryman when 44 kegs of powder were laying within 10 feet of it.

Every person handling powder ought to know there is a three-cornered piece of sheet iron or steel
attached to one end of every keg of powder, which covers a bung hole, and is held in position by a very
small piece of sheet steel bent up behind its broadest end, and by simply bending back this small piece
of sheet steel the three-cornered piece which covers the bung hole can be easily removed and put back
again, after the powder needed has been taken out, with perfect safety.

HOW TO BLAST A SALAMANDER.

By Salamander is meant the mass of chilled iron that is left in the bottom of an iron furnace when it is blown out for repairs. This mass of chilled iron is often to heavy to be moved and must be broken where it lies in the bottom of the furnace stack, and as this is never far from many thousands of dollars' worth of machinery (blowing engines, pumps, etc.), it is very important to use great caution so as not to throw fragments to injure the machinery, and yet break the salamander with as little delay as possible. To do this I recommend using sixty per cent. dynamite.

The reason for drilling so many holes first before firing is, the block is then large and solid, and a steam drill can then be easily set up and operated on it, and as the drilling has to be done before the salamander can be removed, it is best to do it all at once and then remove the drill and blast until it is small enough for removal ; some large pieces may show signs of fracture, and may be reduced by simply laying three or four pounds of dynamite on it and putting two or three shovelfuls of wet sand on top and firing it in the usual way.

First. **All surrounding burned brick** should be dug away and removed, and every part wedged off that can possibly be gotten off, so as to make the mass as small as possible before beginning blasting and to leave the iron unsupported, so that the slightest jar may tell to best advantage on the unsupported mass of iron.

Second. **Then drill a row of holes** one foot apart in a line along the center from its extreme parts (see fig. 47), then drill a hole one foot from every second hole at right angles from the center line of holes (see fig. 47).

Third. **These holes may be any diameter** that can be easiest drilled and should go as near as can be judged to within six inches of going through the salamander. If the iron is very hard to drill and the drilling has to be done by hand, one inch in diameter is a good size ; but by far the best plan to drill holes is to use a steam drill and start the holes at one and one-half inches in diameter and let them reduce in diameter as they go down and the steel wears. The salamander will then have an appearance like fig. 47.

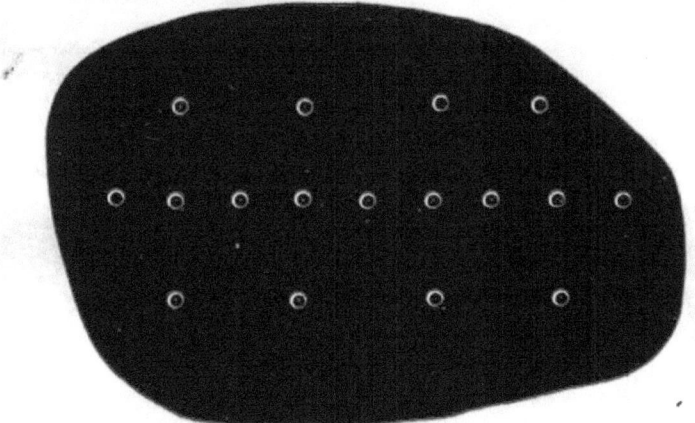

FIG. 47.

Fig. 47 shows a salamander drilled and ready for blasting.

The blasting of a salamander should by all means be done by electricity and all the holes should be charged and fired at once.

HOW TO CHARGE HOLES IN A SALAMANDER.

If the holes are too small in diameter at first shooting to admit half of a full-sized cartridge, then take an electric exploder with wires four feet long, and catch it between finger and thumb of one hand two inches back of the cap, and with the other hand straighten out the wires, taking care to never pull by the cap, then place the cap at bottom of hole and from an open cartridge crumble in a half a stick of ordinary size sixty per cent. dynamite and push it down on the cap with wooden rammer as it goes in, so as to get it all packed at bottom of hole around and on top of cap. But if the hole will admit a half of full-size cartridge, then attach the cap and wires as shown in fig. 18. Then tamp with dry sand in the usual way, and when every hole has been charged then connect the wires, as directed under **Electric Blasting** (page 30), and **fire** as there directed. The first shooting will likely enlarge the holes so that each hole will admit half of a full-sized cartridge without losing time to crumble it into the hole, then attach the exploder as directed in fig. 18, and proceed thus to load and fire all remaining holes four times with half-stick charges. If any holes remain unexploded after the fourth round, then put in three-fourths of a cartridge at a charge and fire three times with that charge, and if any remain then put in a whole stick at a charge in each hole and continue firing and firing until it gives way.

BLACK BLASTING AND SPORTING POWDER

Is made by two different styles of machinery: press cake and stamp mill powder. All first-class powder mills make press cake powder with a plant of machinery costing from $40,000 to $50,000, while a plant to make stamp mill powder seldom costs $5,000. Anyone can know from this that press cake powder must be far superior to stamp mill powder, else why should $45,000 be invested, if a $5,000 plant would do as well? Any person could easily tell whether he is getting best or second quality by selecting a grain of similiar size from each grade and place them on a smooth, hard-wood surface and then place a flat metal surface, such as a knife blade or flat chisel, and then press it down hard; the difference would soon be apparent. The press cake powder being much the hardest, it has, therefore, more explosive material in same space and, of course, will keep much longer without absorbing moisture. The following cuts

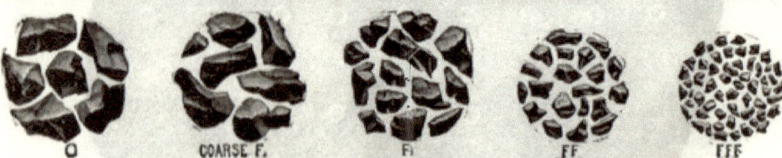

O   COARSE F.   F   FF   FFF

FIG. 48.

show the five different sized grains in general use for blasting and they will enable anyone to order by mail and get exactly the size grain he wants. O and coarse F are seldom used except for very large blasts; single F is generally used for limestone or quarry work, and F and FF for coal blasting.

44

HOW TO USE THE SHOT GUN.

For the benefit of new beginners I print the following instructions on how to use the shot gun :

"Let me tell you how to learn to aim a shot gun. It is a very simple thing when once you have mastered it.

" Lift the weapon with both hands, the right clasping the stock just below the guard, the left supporting the barrels. Look with both eyes steadily at the object to be shot at, and at the same time bring the mid-rib of the barrels straight under the line of vision of the right eye. Pull the trigger instantly.

" Even after you have learned to control your nerves you will find it very hard at first to hit your bird, because you will forget to aim ahead of it if flying across your line of sight, or above if rising, or below it if flying downward.

" In hare shooting it is necessary to "allow" for running by aiming a trifle above the game when it is running straight away from you. This is because your line of sight is above it as you stand.

" The shot gun requires the very best of care in order to do good work. It must be kept perfectly clean and must always be loaded to suit its "habit," as I call it. By this I mean that each gun has a capacity or quality for shooting a certain load best and any other load will lessen its effectiveness. By a little experimenting you can find out the load that best suits your piece.

" Carry your gun on your shoulder with the muzzle elevated and the hammers down, save when you are expecting game to rise, then you may hold it at " ready," which is as follows : Cock both barrels, grasp the stock with the right hand, as in firing, and sustain the barrels at an upward angle in the left hand just in front of and across the breast; the breech-heel a little below the right elbow. This gives perfect freedom of action when the game rises. Moreover, it is the safest position in which to carry the gun, both for yourself and your companions, if you have any.

" Never be in a hurry with a gun, no matter what the apparent emergency—it is the deliberate and cool sportsman that is quickest and surest. Remember what is done as a habit is done perfectly, and all that you have to do to make a crack shot of yourself is to learn to fire habitually by the most approved rule."—Maurice Thompson in *New York World*.

CLUB SPORTING.

This powder is designed especially for breech-loading guns, and is a specialty in gun powder. It will give great penetration, burn moist, and leave the gun in as good condition after a full day's shooting as after the first charge is fired in the morning. Its superior strength and cleanliness, coupled with the low price at which it is offered to the trade, makes it the leading brand of sporting powder now on the market. This powder costs but little more than rifle powder.

 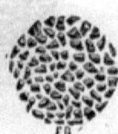 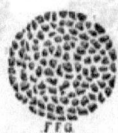 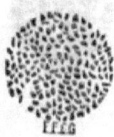

FIG. 49.

Packed in twenty-five pound, twelve and one-half pound and six and one-fourth pound screw-topped metal kegs, and three pound, one pound and one-half pound canisters, all handsomely painted or japanned in blue. Full weight guaranteed in each package.

CHAMPION DUCKING.

Until within a few years this grade of Sporting Powder was put on the market at the same price as that of the highest priced grade of any of the American manufacturers, under the designation of "Diamond Grain." It is now offered to the trade at a price corresponding with that of the second best grade of other well-known manufacturers, and for strength and cleanliness is not excelled by any sporting powder at a similar cost now on the market.

FIG. 50.

Packed twenty-five pound, twelve and one-half pound and six and one-fourth pound screw-topped metal kegs, and three pound, one pound and one-half pound canisters handsomely japanned in vermilion, and covered with heavy paper wrappers. Full weight guaranteed in each package.

RIFLE POWDER.

  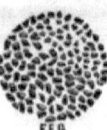  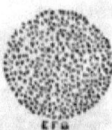

FIG. 51.

Packed in screw-topped metal kegs, painted green, and in screw-topped canisters japanned green. Kegs of twenty-five, twelve and one-half and six and one-quarter pounds. Canisters of one pound and one-half pound—twenty-four canisters in a case. Full weight guaranteed in each package.

SPORTING AND BLASTING POWDER.

"CRACK SHOT."

This new brand is made from the best selected materials, properly mixed, and for trap and field shooting is A No. 1 powder. Packages painted terra cotta color. Packed only in twenty-five pound, twelve and one-half pound and six and one-quarter pound kegs. Below is given a fac simile of size of grains.

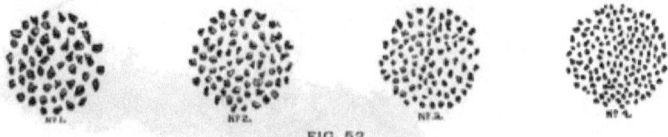

FIG. 52

Be sure to try " Crack Shot." It is guaranteed the best. This brand of powder has been highly recommended by trap shooters and many declare it is the best powder for all sporting purposes they have ever used.

SMOKELESS POWDER.

Much has lately appeared in newspapers about smokeless powder, and I was last fall invited to witness experiments in stone blasting in a quarry with smokeless powder. The experiments I witnessed were very far from being satisfactory to me, and when the newspapers came out next day with a glowing statement about the experiment I had seen, totally at variance with the facts of the experiments, I concluded, so far, the whole of these experiments was a farce, and I have not heard of that make of powder being used in practical quarry works, although praised so highly in the newspapers. Common black blasting powder and dynamite are now so cheap and good that I have no hope of ever seeing a smokeless powder so cheap that it can be used in competition with the explosives now in use in quarries and all open air blasting, and smoke is so little inconvenience in such work that I do not think it worth space in this book to say any more about it.

MY EXPERIENCE WITH DYNAMITE AT JOHNSTOWN, PA., IN 1889,

Leaves no doubt of its great superiority for removing obstructions in streams over any other known plan whether it be removing natural rock under water, or breaking a jam of logs, ice or debris, or removing

FIG. 53.

Johnstown jam above the P. R. R. stone bridge, taken on the fourth day after the flood of May 31, 1889, from a point about two hundred feet above the bridge on the west side of the stream, showing Pittsburgh firemen throwing water on burning jam above the bridge.

sunken coal or steamboat wrecks. Anything of this kind can be cheaply and quickly removed by a judicious use of dynamite, but perhaps a brief narrative of what I saw and did at Johnstown will give a better idea of its use than any other way I can state it. See fig. 53 for what I saw when I arrived at stone bridge of the Pennsylvania Railroad at Johnstown, about noon of the first Monday after the flood (which had been the previous Friday). I found such a combination as was never seen in the world before. The South Fork dam had broken about 3 P. M. on the previous Friday and a mighty torrent of water, seventy-five feet high and about two hundred and fifty feet wide, rushed through the opening in the dam, and rushed on and on at a furious rate for nine miles to Johnstown, carrying death and destruction to everything in its mad career; locomotives were swept along on the crest of the waves as if they had been sticks of cord wood, and car loads of pig metal, weighing at least forty tons, struck and instantly demolished strong brick houses over one mile from the railroad track they were standing on before the flood picked them up, and a quarter of a mile from any railroad track. On their route they must have crossed the Conemaugh river, which at that instant must have been at least sixty feet deep.

It has always been a matter of wonder to me how it came that the water, after traversing the nine miles from South Fork dam to Johnstown and filling up all the valley for that distance, yet everyone who saw the flood strike Johnstown, described it as a wave thirty feet high, traveling at great speed and wrecking everything in its way. This was recently explained to me by an eye witness. The water was near one hundred feet deep, and was one mile wide and three miles long in the dam before it broke. All of a sudden two hundred feet of the dam gave way and this huge volume of water rushed down the valley, which was from three hundred to five hundred feet wide, with high mountains on each side, and this valley, almost through its entire length and width, was a dense thicket of healthy growing young trees, principally young maples, birch and willows. When the torrent struck these trees it tore them all out, roots, earth and all, and rolled them before it like a rolling, movable dam until it struck the viaduct stone bridge four miles from where it started. This was a model stone bridge about seventy feet high with single semi-circular arch of eighty feet span of remarkably good workmanship and with embankment was about two hundred feet long, double tracked for the main line of the Pennsylvania railroad. The above described mass of rolling debris was thrown with great force against this bridge and in one minute formed a dam near one hundred feet high, and then the bridge gave way and was practically the breaking of another dam. The flood, having been held back until its straggling forces came to the front, the bridge gave way and the huge flood, freed from its confinement, burst with renewed force again on its mad career towards Johnstown. About two miles above Johnstown it was again stopped by another stone bridge of about the same dimensions as the first one. This again held back the advanced column and piled up an immense volume of water until the second bridge gave way and the great flood, having rested to collect its forces, burst with unresistable force on Johnstown. All this time it was accumulating debris of every detcription, which was finally arrested by the stone bridge in Johnstown and formed such a mass that nothing but dynamite could remove it.

It can thus be easily seen that this mighty torrent, during its nine-mile journey, which it is estimated was made in about one hour, picked up huge trees as well as small saplings by the roots, earth and all, by the acre, and as it passed farm houses, and the villages of Mineral Point and Woodville, and swept at lightning speed through the densely populated and business part of Johnstown, it must have had on its surface dwelling houses, stables, fences and wood piles by the acre. But fancy this mighty cavalcade suddenly arrested and its first part brought instantly to a standstill by the strong, substantial, four-track stone bridge of the Pennsylvannia Railroad, having six very strong piers and seven stone arches. The advance part of this debris soon formed a dam above the bridge, so that the water rushed over the top and across the bridge several feet deep. The light floating houses, etc., soon stuck on the bridges, and others, coming on with great momentum, jammed tight against those before them. Thus several acres must have been packed tight and close with floating frame dwelling houses, stables, fences, etc., floating high and dry on top of the water, when very soon

THE WHOLE MASS CAUGHT FIRE

And had burned all Friday night, Saturday and Saturday night, Sunday and Sunday night, and had the appearance as you see on page 48 when I saw it at noon on the first Monday, or three days after the flood. It must have been a very hot fire, for large, heavy, green trees, which had been caught in a standing position, had been burned off as if they had been old dry fence rails, and all had burned down to the water's edge, yet when measured a few days after the jam it was

350 FEET WIDE BY 850 FEET LONG,

And estimated to be about fifteen feet deep on the average, composed, of course, of the heaviest part of the floating articles, such as very large stumps of trees whose tops had been burned off by the fire, gondola cars, beds and trucks, long stretches of railroad rails with ties still attached, very heavy bridge beams with what appeared to be hundreds of miles of telegraph wires and heavy copper electric light wire; this wire was the cause of great trouble. Now when the reader reflects that the flood of water had run over this mat of heavy material for at least forty-eight hours and being retarded in its speed, dropped **thousands of tons of sand, stone and gravel** into the interstices of this mat, and when he remembers that this mat or jam was 350 feet wide by 850 feet long and average depth fifteen feet (see fig. 54), he can have a faint idea of the difficulty of removing it; and here I may remark in brief that **every other known means for removing it was tried, but without success.** Some two days were lost in getting dynamite, a battery and assistants together, and as the proper place to commence operation was at the down-stream edge of the jam, so that every piece set free would be at once carried by the current out of the way of further operations, yet as all the material here was piled up tight against the piers of the stone bridge, and for fear of injuring the bridge, only small charges of dynamite could be used until the blasting could be done farther away from the bridge. All this and the novelty of the work caused the work to go slow the first week, and great impatience was expressed at the slow speed made by dynamite and a great many plans were proposed for removing

THE GREAT JOHNSTOWN DRIFT JAM MORE QUICKLY.

First a wrecking train of the P. R. R. was stationed on the stone bridge (see fig. 54), but in less than one day the whole gang was disgusted with their progress and quit. Then a Mr. Colburn, a contractor from Altoona, insisted to Gen. Hastings that he could take the whole drift out in a few days, and I was requested to suspend dynamite blasting ; but in less than forty-eight hours he left in disgust ; then some twenty steam engines were set at work and several of the largest patterns carrying steam up to one hundred and twenty-five pounds were placed on the bridge and two days ended their work, for no cable could stand the strain of dragging even what appeared to be very small pieces out of the pile until it

FIG. 54

Johnstown jam above the P. R. R. stone bridge, taken eight days after the flood, about two hundred feet above the bridge on the east side of the stream.

51

was shaken up by dynamite. Then Col. Biggler, from Clearfield county, Pa., insisted that if Gen. Hastings would give him authority he would bring down three hundred men from Clearfield county, experienced loggers, and they would make short work of removing the jam. By and by they came and did jump into work lively, but one day's work showed them the utter worthlessness of cant hooks and pike poles until the mass was shaken up by dynamite, and I was again called upon to resume blasting. I only mention these facts to show the great superiority of dynamite blasting over every other known plan for doing such work, for it was no guess, but by actual test every other known plan was tried, and that thoroughly, for every one who had boasted that he could do it naturally did his best before abandoning it, but after every other plan had been tested to its utmost it was left for dynamite to finish the job. Perhaps it may be interesting and useful for the reader to know another use of dynamite at Johnstown. This was

FIG. 55.

Johnstown jam above the P. R. R. stone bridge, taken the sixth day after the flood of May 31, 1889, from the west side of the stream about five hundred feet above the stone bridge, showing the jam before much had been done to remove it and showing the trestle work, across washout at the east end of the stone bridge, where the flood in a few minutes washed away four tracks of railroad and an embankment forty feet high and rushed through into the stock yard of the Cambria Iron Company

BLASTING DOWN A LARGE BRICK ROMAN CATHOLIC CHURCH.

Amid such a disturbance among dwelling houses, as I have already described, it was quite natural that some of them should catch fire from the upsetting of cooking stoves, etc., and as the burning build ings floated along, the fire quickly communicated to other buildings, among which was a Roman Catholic Church, which would have stood the flood all right if it had not caught fire from the passing burning houses. This burned out every particle of wood about the whole building, outside and inside, and nothing but high, naked, brick walls were left standing, all of which had been much loosened by the fire, and as thousands of people were passing near them day and night, it was ordered to be blasted down; this was done by Thomas McNally, who had thirty-three one inch and a quarter holes drilled in

FIG. 56.

Scene of Johnstown jam taken two months after the flood, from a point five hundred feet above P. R. R. stone bridge on the west side of the stream showing the great Johnstown jam all removed stream cleared of all obstructions and men at work repairing the bridge.

the brick walls at regular intervals around the building about two feet above ground, and had all charged and connected to a battery, and all fired at once. The walls gave a sudden shudder and just settled down a shapeless mass of bricks and mortar, and all this was done without throwing ten pounds of anything fifty feet from the building. I am thus particular to give details of this blasting, as I think it a plan which could often be used to good advantage in removing dangerous walls or chimneys left standing after a fire. The details of

HOW THIS BLASTING IN THE JAM WAS DONE

will, no doubt, be interesting to my readers. At first, owing to the newness of the work, I naturally had to try different experiments, until I finally settled down to about the following routine: I had from eight to twelve assistants, and they were divided as follows: One in charge of blasting apparatus, one in charge of stock of explosives, two making torpedoes and the balance were with me on the drift. The great bulk of our shooting was done with charges made as follows: First, we would get a piece of flooring or fence board, one by six inches wide and ten to sixteen feet long, then hunt over the jam for nails out of burned buildings, then drive three or four eight or ten-penny nails through the board, six to eight inches from one end, then tie as many sticks of dynamite (one inch and a quarter in diameter and eight inches long) around among the nails. I found it always necessary to have the nails project half length on each side of the board so as to prevent the board pulling out of the bundle of dynamite sticks; then do the same with another series of nails driven half through the board, and another jacket of sticks of dynamite laid lengthwise of the board and tied tight with two tyings, one at each end of the dynamite cartridges; and thus I put on as much dynamite as I thought necessary in each place, and then I put two electric exploders in each charge, placed one in each end, and one wire of each exploder fastened to one wire of the other exploder; then I would fasten the two remaining wires to top of board. While a torpedo was thus being made, I had a man removing small pieces of wood, stone, etc., so as to form a receptacle or well for the torpedoes as deep under the water as possible; then I would prepare three or four of such around a sunken mass and fire them all at once with a battery, and the result would be as shown in fig. 58.

FIG. 57.

Johnstown jam above the P. R. R. stone bridge seven days after the flood, taken from the west side of the stream about five hundred feet above the stone bridge, showing a beginning made of removing the great jam under middle arches of the stone bridge.

If I remember right, after deducting the time lost by the various stoppages above referred to, I spent about twenty-five days in actual blasting on the great Johnstown jam. At least five of these were practically lost at starting the work, for want of experience in such work and suitable tools, as I had first to find out what tools I needed and then send to Pittsburgh for them, for all the hardware stores in Johnstown, with their stock, had been destroyed. **Thus in twenty days actual work** I removed the unprecedented Johnstown jam and changed the appearance of the stream from that shown in fig. 57 to that shown in fig. 56.

DYNAMITE EXPLOSION AT THE BRIDGE.

FIG. 58.

BEST METHOD OF SINKING A SHAFT OR RUNNING A TUNNEL.

Much money is now invested every year (or rather wasted should be the word) for the purpose of obtaining valuable minerals by sinking shafts, etc. Much of this might be more profitably spent, if the owners would adopt the Davy Crockett rule to "first be sure you are right, then go ahead." Many shafts and slopes or tunnelled openings that I have known of have been commenced and carried on, apparently on the principal that any person can make a hole in the ground, which, of course, is true in a sense, but in this, as in many cases, " wisdom is profitable to direct." And, although shaft sinking is not quarry work, yet it is a kindred subject with which I am somewhat conversant and on which I will volunteer a few remarks. It frequently happens that owners are so anxious to get at the hidden treasure that they underestimate the difficulties of the undertaking at the beginning, but they are generally better informed before the job is finished, and the result of my observation is that no person should decide to sink a shaft unless he can easily command at least fifty per cent. more money .than the estimated cost of the undertaking, with a reserve fund of fifty per cent. more to fall back on in case of big unforseen expenses, for when once commenced, the quicker it is completed the better, no matter what difficulties may be encountered " to the bottom or China" should be the rule. Among unforseen expenses I may mention an unexpected trouble from water, from labor troubles, breakage of machinery, unexpected hardness of rock, etc., and first I would here remark that first and foremost, it is absolutely necessary that a man of strictly sober habits be put in charge of the work. From my observation I say it is utterly useless to expect satisfactory results when the man in charge must get a drink every time he gets wet, and another every time he is dry ; in fact he should be a strict teetotaler and a man of experience in the business of shaft sinking. Without such qualifications, it is utterly useless to expect a satisfactory job.

Second, he should be a man thoroughly conversant with the latest improvements in his business. A man who was a good shaft sinker ten years ago, but has been out of the business ever since, practically knows nothing about shaft sinking now. The introduction of air compressors and power drills, electric blasting, dynamite, and improved hoisting engines, improved pumps for taking away water—these and many other causes have completely changed shaft sinking, and a man who is ignorant of any and all of these, or is opposed to the use of any of these, shows he is not up with modern improvements and is utterly unfit to be superintendent of sinking a shaft, for I have known of men put in charge of work who even condemned and had modern appliances condemned and rejected, when it was plain to any one who understood the business that the whole trouble was the modern improvements would finish the job too soon and the superintendent would be out of a situation. But the worst trouble to be expected is from water —especially so, if the shaft is sunk at the base or bottom of clinical formation, that is, at a depression or dip in the geological strata. The water will then pour in from all sides and continue to do so until all the receptacles of water in the region have been exhausted. This, of course, makes the trouble worse while sinking the shaft, but if you want a shaft sunk and have settled on the spot, then you must prosecute it vigorously and if possible permit nothing to interfere or stop work (of course excepting Sabbath Day), until you go as deep as needed, and as a large amount of water may be expected at any moment, the first thing to be done is to have a great surplus of

PUMPING POWER,

Which should consist, first, of plenty of steam boiler capacity (if fuel is cheap). This should consist of a large amount of cylinder boiler power, because the water you pump out of a shaft will likely have to be used in the boiler, and it may be expected to cause heavy deposits of mineral, particularly in the boiler, which may, before you suspect trouble, burn out a sheet or two over the fire and cause a stoppage of pumps and the filling up of the mine again. This should by all means be guarded against. But if fuel is scarce and expensive, then tubular boiler should be used, the same as shown in fig. 63, which is the most economical steam-producing boiler I know of for that use and at the same time easily repaired. Third. But with any form of boilers using water from a mine a liberal quantity of anti-incrustation preparation should be used. One pint per day of pure, natural petroleum is said to be the best preventive and should be freely used. Fourth. The steam boiler power should be divided into two batteries and be used alternately, so as to give plenty of time for cleaning, inspection, repairs, etc. Fifth.

THE STEAM SHOULD FIRST BE USED THROUGH AN AIR COMPRESSOR,

And the compressed air used on the pumps. This will be found by far the most economical and satisfactory mode of using steam to run pumps under ground, and is being rapidly introduced for this purpose and should by all means be used for shaft sinking, so that in case some unavoidable accident occurs by which the work becomes flooded (if the pumps are of the duplex pattern), there will be no trouble about starting them under water and making them pump themselves out again, but if steam is used direct on the pumps it will be impossible to start them under water, because the steam pipe and steam cylinder, being exposed to such cold water, would act like so many feet of condensing surface to kill all the steam that could be thrown into them, but with compressed air this could not be felt. Another great trouble with steam in a shaft is that the hot steam pipes, passing down to the pumps, keep the whole shaft hot and murky and drop hot water from the pipes on the men below, and as the pumps have often to be lowered and the pipes lengthened when steam is used, they are hot and very hard to handle by the workman, whose hands are sensitive to heat.

BUT WHEN COMPRESSED AIR IS USED

All this is changed; every part of pipe or pump is cool and no trouble to handle. Every escapement from the pump is good, pure, fresh air, same as the atmosphere above, but has been cooled in the air compresser, and this is of great advantage by expelling the offensive gases after blasting, which enables the men to go down sooner after every blast, and when down they can work with more vigor. Besides, the

COMPRESSED AIR CAN BE USED TO

Run one or more steam drills, the advantage of which can be easily seen by reading what is said about steam drills on page 62, where it is apparent that one man with a steam drill can do more work than twenty men. It therefore amounts to putting in twenty men, which it would be impossible to work in a shaft, and after the holes are drilled then

57

USE DYNAMITE, AFTER TRYING SEVERAL GRADES,

To find out the grade that will do your work best (see page 19). Then charge the holes heavy, especially center holes, as dynamite is now so cheap you will find it much cheaper and quicker to have plenty of power-made holes to blast and reduce your rock to shovel stuff with dynamite than by sledging. In drilling down through some of the shale strata, the borings may be found of such a fine adhesive character, that it may adhere strongly to wings of power drill and prevent the drill from operating properly. I have known cases where power drills have been condemned and thrown out because they stuck in the mud in the hole, but that was only a pretense. The real fact was that they did not put in enough water to drown the dust and wash it out as soon as made, but wanted the drill condemned and do the work by hand, so as to lengthen their job, and, of course, all this increased cost to owner. But you can get over all that trouble by reading what is said about a water injector on page 65, while the advantage of using such a plant as I have described above is clearly seen by reading an extract from a letter sent me by Mr. J. K. Taggart, formerly superintendent at Leisenring Mines, Fayette County, Pa.

A. Kirk & Son:

GENTS:— I used your drill to put down fifteen holes, each six feet deep, in the bottom of our large shaft, and charged all with your dynamite, and loaded all to fire with your battery. We made the connection, we thought, as you directed, but it would not fire. Upon inspecting the wires, as you direct in case of misfire, we found that the last man to come up had caught his foot in the wire, as he got into the bucket, and broke a splice. This was repaired and the battery operated again, when the whole fifteen holes were fired simultaneously. We hoisted **one hundred and eighty-five buckets,** each containing about three-quarters of a yard of rocks, before we needed to fire again.

Yours Respectfully,

J. K. TAGGART.

THEN USE ELECTRIC BLASTING.

For instructions see page 30 and you will find by experience that to use fuse for blasting in shaft or tunnel is just throwing money away, delaying the work and endangering lives of men, for if there is danger in the world it is to go down on a hang-fire shot, because the fuse has missed fire and a man cannot tell whether it has been drowned or is liable to go off any moment and without fail kill every one in the shaft; but this can never occur when the holes have been charged to fire by electricity. See page 32.

PUMPS

Are now an indispensible part of a contractor's outfit, and it is very desirable that a pump be reliable— that it is not liable to go out of order by openings, chokings or valves breaking, and the fewer parts there are in a pump, or any machine, the less risk of going out of order. After considerable experience with pumps, I have found the Hall Steam Pump to come the nearest to perfection for ordinary suction and forcing purposes, as will be seen by referring to their advertisement on back of cover.

FIG. 59.

DUPLEX PUMPS USED IN SHAFT SINKING.

Pumps for shaft sinking must have as much pumping capacity as it is possible to have with the same weight of metal. After considerable experience with pumps, I am confident the Hall steam pump comes nearest the above requirements of any pump I know, and, besides, should the water be pumped dry out of the sump, it makes no practical difference in speed of pump. By a peculiar arrangement of valves it will only run just so fast and no faster, although pumping nothing but air.

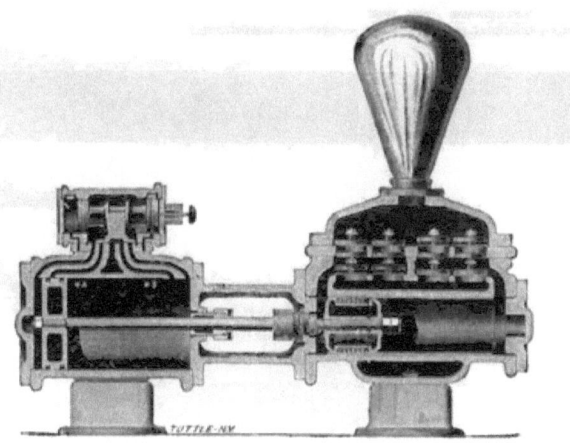

FIG. 60.

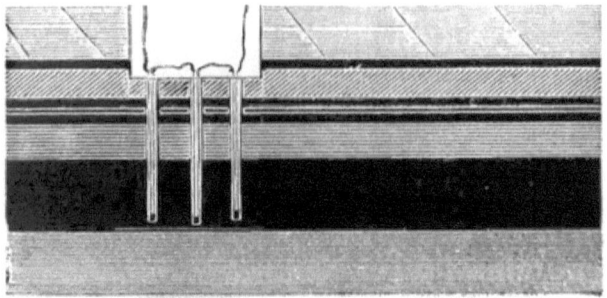

FIG. 61.

The above cut represents a blast of three holes, each six feet deep, in the bottom of a shaft which may be many hundreds of feet deep. As soon as the operator in the shelter house pushes down the operating rod in the battery, as described on page 31, all the holes on the circuit (of any number up to 35) will explode simultaneously and the result will be as shown on next page, where anyone can see at a glance that the shaft has been deepened six feet at that spot and the muck has been broken up fine enough for easy removal.

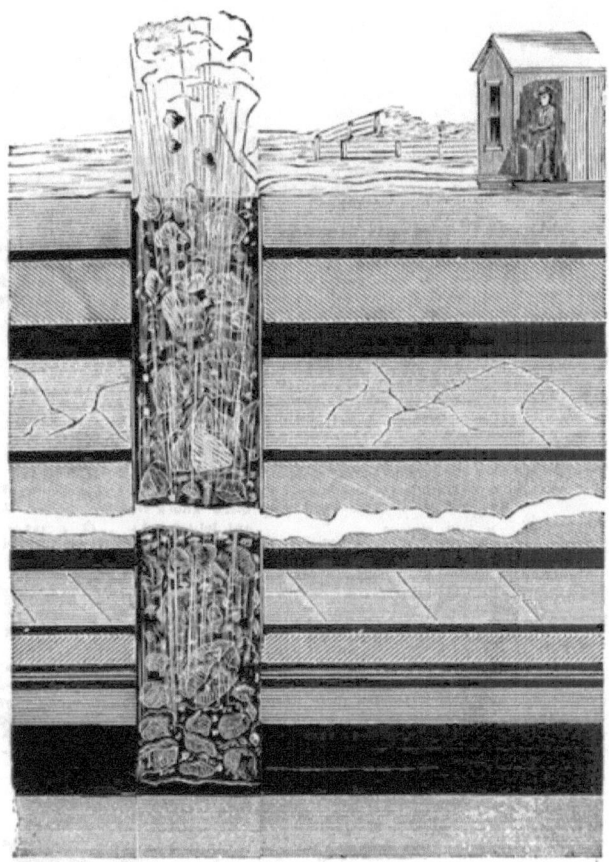

FIG. 62.

By using a steam drill to make fifteen holes, each six feet deep, in the bottom of a large shaft at Lesingring, and using dynamite and electric blasting, Mr. Taggart, superintendent, assured me that he removed one hundred and eighty-five bucketfuls of muck material from one blast, and that was done as quick as the buckets could go up and down, as the stuff was so broken and easy to load.

SHAFT SINKING.

Being anxious to give my readers all reliable information on the subject of shaft sinking, I submitted a proof of the foregoing article to Mr. William Allison, of Cresson, Pa., who has had many years of practical shaft-sinking experience, and below I give his answer, which contains much valuable information which I am sure will be profitable to my readers :

Arthur Kirk, Esq., Pittsburg, Pa.
CRESSON, PA., Aug. 21, 1891.

DEAR SIR :—Yours of Aug. 13th received, also copy of the notes on shaft sinking. I have read them over very carefully and I think it very good, as far as you apparently want to write on the matter of shaft sinking. I have a few items I wish to mention if you think them proper. First, in speaking of boilers, I would say that cylindrical should be used in all cases where mine water is used, regardless of the price of fuel. You can put what you have a mind to in a boiler and the best it will do is prevent the mud from getting hard and sticking to the shell. What is wanted about a coal mine, where mine water is used in a boiler, is that a man can get into the boiler and scrape the mud out clean, which he cannot do in a tubular boiler, besides he must be able to clean it properly. There is a great amount of mud which collects between the flues, which is impossible for a man to get out from either top or bottom. Manhole flues being put in a boiler, as a general rule, contracts the space so that a man cannot even get an iron rod pushed in between. I have seen a great deal of those so-called preventatives used and I have come to the conclusion that in a great many cases the preventative is worse than the disease, but as you say, petroleum is the best, I also believe it is the best. The next thing I want to call your attention to is the use of compressed air in pumps. The advantage of compressed air over steam is so great that a person with a little experience with both steam and compressed air could write for a week on the matter and not exhaust the subject. In the first place you have nearly the same pressure on your pump piston down in the mine that you have in your receiver on top of the ground, no matter how great the distance be. In the second place you have no condensed water in your pipes which prevents your pump from working, besides the loss of steam pressure caused by the above condensed water. In the third place you do not have the exhaust steam from your pump making the shaft so hot that it is almost impossible for men to live in it, unless it is properly partitioned off, which is a very hard matter to keep in good repair in a sinking shaft, as the blasts will very often knock your boards off. With compressed air your pump is ventilating the shaft and there is no lost power. As for the power drill, I would say this much, when using steam the drill is so hot that men can not catch hold of it, and it takes three or four men to move the drill when hot, because the only place the men can catch hold of it is by the tripod. If compressed air was used one man could move it in many cases, and then if using steam you have the exhaust of the drill to contend with. Between the exhaust and the heat of the drill, things are pretty hot, as a rule, and the time lost in repairing hose which burn out at the ends is very great.

Hand drilling is not to be compared with power drilling. The difference is so great in favor of the power drill that one trial will convince any practical man, but the man in charge will, as a rule, have some trouble in introducing power drills. The workman will try everything to make them a failure. It is therefore necessary for him to have all orders carried out to the point. When the men see the good work the drill can do and how it lessons their labor, they will approve of it and everything will go smoothly. The following figures show a test made in the Cresson shaft in gray limestone, very hard, on the 11th of February, 1888, power drill test, one man running drill:

	POWER DRILL.				HAND DRILL.			
First	Hole,	4 feet 9 inches deep,	29 minutes	First	Hole,	3 feet 6 inches deep,	70 minutes
Second	"	2 " 8 " "	10 "	Second	"	4 " 6 " "	90 "
Third	"	2 " 8 " "	11 "	Third	"	2 " 6 " "	50 "
Fourth	"	5 " 6 " "	35 "	Fourth	"	4 " 6 " "	80 "
		15 7	85			15 0	290

The drill should be supplied with plenty of water to keep the dust wet like thin mud and clean properly every time the drill is changed, but the mud in the hole is not to blame at all times for the drill sticking. It often happens that the drill gets out of line if mounted on a tripod. The weight and motion of the drill when working causes the legs of the drill to sink in the soft slate, therefore the drill gets to one side and gets to bear on the side of the hole and becomes fast. To overcome the above trouble is very simple: get three pieces of 2-inch plank, 10 or 12 inches square, place one under each leg and it will keep in good working shape. In speaking of dynamite, you recommend heavy charges in the holes. I wish to say that great care has to be taken not to charge holes *too heavy*. Over-charging might clear all the timber out of the shaft. The pump is always from 6 to 25 feet from the bottom, and an over-charge might break it up or move the timbers from under it and let it fall. One more word about boilers before I close. I think when coal companies are getting boilers for use at their mines where coal is to be had at first cost, that the point in view should be to economize in boilers. To do so they must get cylinder boilers; next choice would be two large flue boilers. Away from coal mines, where coal will cost four times more than at the mine, then it is best to economize in fuel. To do so they must get return tubular boilers. Hoping what I have written will be of some good to you, I am

Yours respectfully,
WILLIAM ALLISON.

MACHINERY DEPARTMENT OF QUARRY WORK.

Steam boilers, steam engines, steam or air rock drills, pop hole drills, hand coal drills, hand fire clay drills, coal miners' drills, scrapers and needles, air compressors (steam actuated), air compressors (water actuated), stone crushers, quarry bars for drilling rows of parallel holes in rock, channel bars for cutting out smooth-faced stone, electric blasting apparatus, water injector to convey water to bottom of hole while drilling, stump augers, dynamite thawing kettles, belt conveyers for conveying crushed stone from crusher, steam pumps of all sizes and uses, wire rope conveyers, etc., etc., all of which it is important to obtain. All machinery should be the

VERY BEST QUALITY FOR THE WORK

The purchaser wants it for. And here I wish to call the attention of persons thinking of purchasing machinery to the importance of getting the very best of any kind. It is a very good rule, and I may say that the latest pattern is the best. This is especially true of rock drills, air compressors, electric blasting apparatus, etc. During the twenty years I have been selling these articles, there has been a constant improvement by the addition of some new device, so that the machines to be had now are far superior to those of fifteen 'or twenty years ago, and farther that a low-priced machine is not always a profitable machine, but it is rather an evidence to the contrary, for a person regularly in the business knows the merits and demerits of his own and his competitors' machine, and if a manufacturer is willing to take less than he knows his neighbor will sell at, he thereby confesses that his machine is inferior to that of others in the trade. While the man who has spent great labor and money to perfect his machine knows he has the best and is more independent in his prices.

OF STEAM BOILERS.

There are so many different plans and makes of steam boilers and every plan and make of boiler has its friends and advocates, so that after forty years' experience and careful study of steam boilers, and all connected with them, I do not know of any one make of steam boilers that can be set forward as suitable for every plant or use. For instance, after supplying a first-class new three-inch tubular locomotive fire-box boiler, which gave best satisfaction the first thirty days and after that gave great trouble by leaking around the end of tubes in fire-box, and after examination by good boiler makers that boiler was condemned and taken away, and another new fifty horse power boiler of same make put in, and it worked all right for about the same time and then commenced leaking around the end of flues in the fire-box. This second boiler was also examined by first-class boiler makers and condemned. They said the holes in the tube sheet of both boilers had been cut about one-eighth of an inch too large and the tubes had been expanded far beyond their natural size to fill up the holes, as they said could be plainly felt by feeling with the finger inside the tube, that there was a ridge all around the inside of end of tube, which the inspector insisted had been done by using the expander too much, and which had weakened the tubes so much that they leaked. This all looked very reasonable until a file was tried on this inside ridge in end of tube, and then it was shown that this hard ridge was nothing but a hard deposit of mineral matter around the inside of every tube in the lower row, and on opening the boiler the tubes were found coated all over with a mineral deposit nearly a quarter of an inch thick, showing that bad water had been the cause of all the trouble and that there was nothing wrong with either of the boilers. I mention this case to show that it is necessary to take in all the surrounding circumstances, and especially the character of water to be used in it, and as all water near quarries is likely to be highly impregnated with mineral matter, which is very injurious to steam boilers, it is of great importance to guard against

INCRUSTATION.

For the usual incrustation in boilers, all that is required to detach and break up the scale is to use Caustic Soda or Concentrated Lye—about one-quarter pound per horse power, or where the Caustic cannot be readily obtained, fifty per cent. additional of Sal Soda, dissolved into the feed water and pumped into the boiler. Tannic Acid, clear extract of tan bark, or even the spent liquor, if settled and made clear, may be used to advantage, a few gallons at a time, for softing and breaking up the scale, and if used in small quantities twice a week will keep the incrustation under control. For facilitating the use of anti-incrustating fluids, an inlet to the suction pipe near the pump, with a cock or valve, should be so arranged as to draw the fluids from a pail or tub. And here it should be stated it is very important to provide a large, easily-operated blowoff cock, which should be kept in first-class order and opened at least every evening and the boiled down water allowed to escape. And where there is any objectionable matter in the water, a suitable reservoir should be provided high enough so that water can be given time to settle before being put into boilers, and every boiler should be blown out empty every Saturday night after drawing the fire, and there should be a large supply pipe from settling reservoir so that the boilers can be quickly filled again with pure water.

STATIONARY TUBULAR BOILERS.

FIG 63.

Stationary Tubular Boilers.

SPECIFICATIONS.

HORSE POWER	10	15	20	22	25	30	35	40	45	50	55	60	70	80	90	100
Diameter of Boiler, in inches	30	36	36	40	42	44	44	48	48	54	54	60	60	66	66	66
Thickness of Shell	…	…	…	…	…	…	…	…	…	…	…	…	…	…	…	…
Length of Tubes, in feet	8	8	10	10	10	10	12	12	14	12	14	12	14	12	14	16
Number of Tubes (3 inches diameter)	13	28	28	34	38	40	40	48	48	60	60	72	72	100	100	100
Square Feet Heating Surface	150	226	300	328	375	440	535	616	676	760	830	914	1,066	1,225	1,340	1,500
Height of Dome, in inches	20	22	22	24	24	24	24	28	28	30	30	34	34	36	36	36
Diameter of Dome, in inches	18	20	20	22	22	24	24	28	28	30	30	34	34	36	36	36
Height of Smoke Stack, in feet	28	28	30	30	30	30	40	40	50	40	50	40	50	40	50	60
Diameter of Smoke Stack, in inches	14	16	16	18	20	22	22	24	24	26	26	28	28	34	34	34
Weight of Boiler, about	2,000	2,350	2,850	3,425	3,850	4,150	4,700	6,150	6,700	7,560	8,000	8,950	9,400	10,600	11,600	12,600
Weight of Fixtures, about	1,200	1,450	1,750	2,075	2,250	2,350	2,500	3,250	3,500	3,500	3,700	3,800	4,200	4,800	5,000	5,200
Weight of Boiler and Fixtures, about	3,200	3,800	4,600	5,500	6,100	6,500	7,200	9,400	10,200	11,100	11,700	12,700	13,600	15,400	16,600	17,800

HORIZONTAL TUBULAR BOILERS.

FULL FRONT.—FIG. 64.

I DO NOT RECOMMEND FANCY BOILERS.

Much has lately been said in public print in favor of fancy heat-saving boilers, but as they all require a skilled man in charge, who would naturally expect more wages than is generally paid about quarry work, in view of this and the trouble of having repairs made at a quarry or contract work, I decline to recommend any of them where stoppage for repairs may cause so much loss, no matter if they may save a few bushels of coal per day.

Engineers and steam users are well aware of the expense, inconvenience and loss of time occasioned by frequent stoppages to repair boilers. These repairs are made necessary by one of three causes. 1st. The *cracking* of the *circle seams* back of the bridge wall, caused by feed water of a low temperature coming in contact with the heated plates. 2d. By the *burning* and *bagging* of the *plates immediately over the grate bars,* occasioned in the old way of feeding boilers by the current of the feed water passing from the back to the forward end of the boiler, carrying with it the loose scales and collecting them over the fire plates, where there is the greatest evaporation of water. Here, by the deposit of the foreign substances held in solution in the water, they soon become cemented together to the bottom of the boiler, excluding the water from the iron, which becomes heated and is forced down by the pressure in the boiler, forming the bulge or bag. 3d. The oxidation of the inside of the stand pipe or mud drum, caused by the action of certain impurities in the water when at a low temperature.

The use of Ford's Patent Funnel Feed Water Heater, near the front end of boiler, will preven' all of these troubles.

The heater being located in the *steam space* of the boiler, the water falls in a thin sheet from the rim of the funnel *B* (fig. 66) into the shallow pan *C*, and thence to the water in the boiler. In its descent it is

66

subjected to the action of the steam, and is raised to a temperature equal to that of the water already in the boiler before commingling with it, thus insuring against cracking the plates by the low temperature of the feed water.

By supplying in the *steam space* of the forward end of the boiler, the equalization current is from the front toward the rear end, carrying the scale and sediment *away from the fire* (instead of *toward* it, as in the old way of feeding), and depositing them in the rear end where they can do no harm, leaving the plates directly over the fire free from any accumulation that would cause the bagging, as before described.

I would respectfully refer you to the following well-known firms, selected from many, who have this heater now in use :

Lacy Furnace Co., Oliver Bros. & Phillips, Pittsburg Bessemer Steel Co., Penn Cotton Mills, Carnegie Bros. & Co., Dilworth Bros. & Co., Birmingham Coal Co., Reymer & Bros., Kimberly, Carnes & Co., Pennsylvania Furnace, Graff, Bennett & Co., Smith, Sutton & Co., Park, Bro. & Co., Standard Bolt and Nut Co., Wm. Clark & Co., Hartley & Marshall, Isabella Furnace Co., A. French & Co., Singer, Nimick & Co., Jas. McNeil & Bro., Joshua Rhodes & Co., Liggett Spring and Axle Co., Moorehead, McCleane & Co.

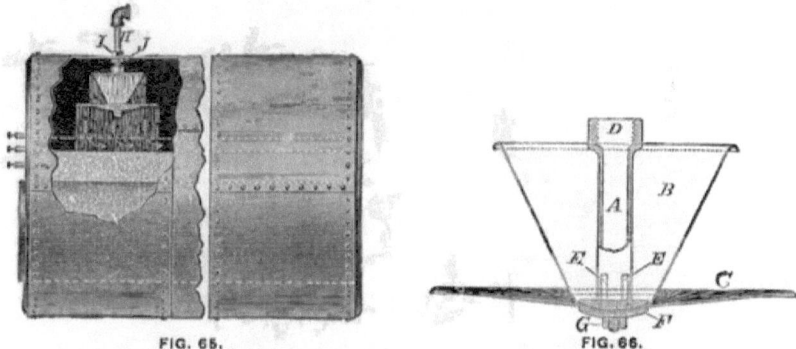

FIG. 65.   FIG. 66.

The two cuts, fig. 67 and 68, show two prevailing types of stationary steam engines. Fig. 67 showing a very powerful, compact, simply-constructed engine, where the balance-wheel is also the driving pulley, and is used extensively where floor space is not scarce ; and fig. 68 shows a powerful centre-crank motion, with duplex cylinders. Very good where space is an object, as on ship board, etc.

FIG. 67.

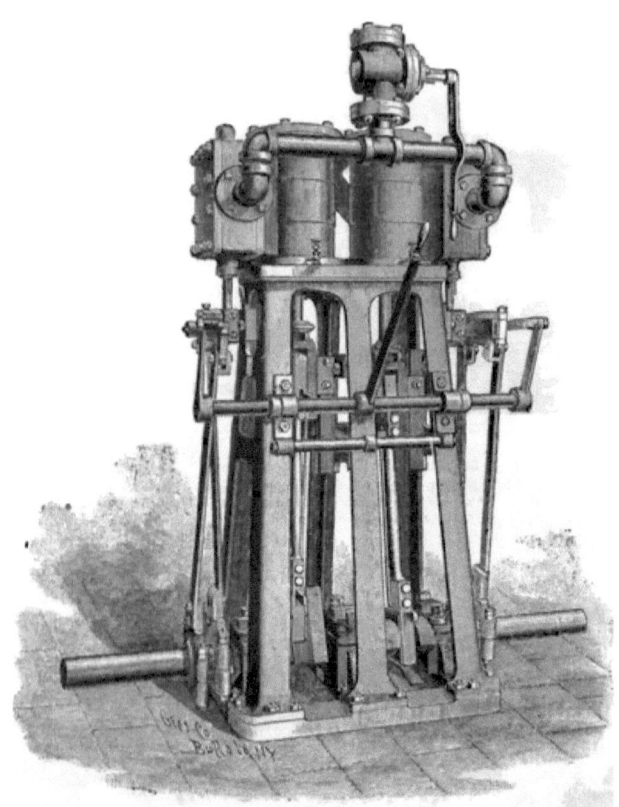

FIG. 68.

AIR COMPRESSORS

Are now admitted to be indispensible to running a stone quarry, or for carrying power to a great distance. Air compressors are now made to be run by steam or water, or electric power, the compressed air from which can be piped many miles without the least perceptible diminution of power, and by using compressed air there is no condensed steam or water to be worked off through the small parts of a steam rock drill, choking and retarding the speed of the drill. But, on the contrary, whenever the throttle cock is opened the **drill moves at once without the stopping,** starting and choking that always attend starting a steam drill after changing steels. Another advantage is, that air escapes quicker than steam, and thus permits the steel to strike the rock a much heavier blow than when the steam above has to spend part of its power forcing the steam out of opposite end of cylinder. Another great advantage of air over steam is, the steam boiler and air compressor can be located far enough from the quarry to be **out of danger from blasting,** and can at same time be located at best place to get coal and water for use with the boilers, and the main air pipe can also be laid away up over the brow of the quarry, and if Ts are inserted every hundred feet in main pipe, then all that is needed is to unscrew a plug from main line and screw in enough one and one-half inch pipe to reach brow of quarry, and then screw on enough hose to reach drill, and drill is ready to start. Air hose costs $20 for fifty feet, while steam hose costs $30 for fifty feet, and at the same time air hose will last four times longer than steam hose. Compressed air is invaluable for pumping mines, as it can be carried any distance under ground to the lowest part of the mine without loss of power, and every escape from the pump

IMPROVES THE AIR OF THE MINE,

And would be of immense advantage many times in thus preventing accidents, such as have recently occurred, and in case of accidents the air pipe, being a strong wrought iron pipe laid on the ground, not likely to be broken by an explosion of gas in a mine, the pump could be made to be stopped or started from the top with its throttle valve in the engine room, and in case of an accident by a heavy fall of roof or the like closing the entrance, the air could at once be turned on from above and good, pure air forced in through the pumps, which should always be a duplex pump, and branch pipes could easily be laid into **bad air places** and thus prevent a collection of dangerous gas, by forcing it out and diluting it with pure air, because movable branches could easily be laid of one and a half-inch pipe away into any part of the mine at very little cost. Compressed air can also be used to good advantage for running ordinary steam engines for operating derricks, rope travelers, hoisters and all such appliances which are located at considerable distance apart, and by using an air compsessor and piping compressed air to each machine, a great number of such machines can be kept running very cheaply by having one large stationary boiler located at safe distance from blasts and conveniently located to get water and coal. In this connection I take great pleasure in calling attention to the lately improved **concentrated piston inlet cold air cylinder compressor,** which dispenses with the old-fashioned inlet air valves with their spiral springs, every one of which was liable to break and make trouble, and what was worse, they took up the space which is now occupied by thick jacket of cold water, which extracts the heat at the point the pressure is the greatest from the air at the very moment when the heat is the greatest, and the whole plan is so strong and substantial that it commends itself to everyone at first sight.

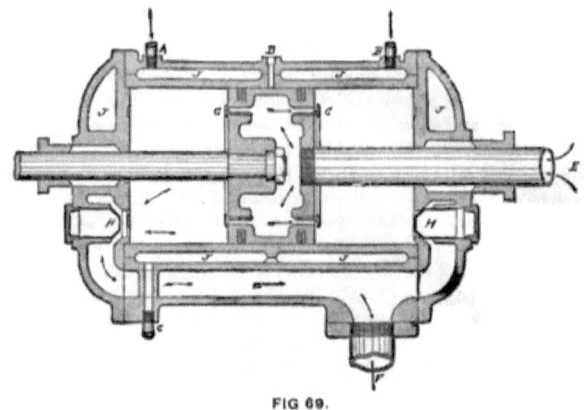

FIG 69.

Fig. 69 shows a longitudinal section of the latest and most improved air compressing cylinder, in which E represents the new plan of admitting air to be compressed. A study of the sectional cut will show that cold water is admitted at A to the space J J J J, which completely encircles the air compressing cylinder and also fills spaces J J in each end of cylinder. The warm water then escapes at B and thus, keeping up a constant circulation, carries off all heat from the air (E, fig. 69) inlet piston attached to the hollow piston head G G and working through a stuffing box in rear end of cylinder. This leaves a large portion of both heads of air cylinders to be covered by water jacket J J, keeping the ends of cylinder perfectly cool, and thereby presenting a cooling surface to the compressed air at the end of the stroke when the air is the hottest. The cold air having been admitted to hollow piston head is fed into each end of the cylinder alternately at E, the valves G G each acting as inlet and check valves alternately, as the motion of the piston head may cause them to open or close.

THE ATTENTION OF ENGINEERS

Is Called to This, the Most Important Improvement in Air Compressing Machinery of this Century.

The special advantages and most important features of piston inlet cold air cylinder are as follows:

1. The free air, before admission to the cylinder, is under thorough control, and may be taken from that point which is the most favorable in its *dryness, reduced temperature and freedom from dust and other foreign matter*.

2. The admission of free air being through a single tube creates a constant and uniform draft of air in *one direction only*, thus filling the cylinder at each stroke with air at *full atmospheric pressure*. In all other compressors the air must be started from a state of rest and put in motion through the inlet valves at each stroke, while here it is *always moving into* the hollow piston, and air being material—that is, having some weight—this uniform movement gives a momentum to the air which causes it *to fill the cylinder*

70

to its fullest extent at each end of stroke. Indicator cards taken on the cylinders of this compressor prove conclusively that not only is the cylinder filled with air at atmospheric pressure, but in some cases the line runs *above the atmospheric line* to the same extent that it runs below it in other air compressors.

3. The air inlet valves are large metallic rings which are not operated by springs, but which open and close by the *natural momentum* given to the valve by the movement of the piston. A study of the sectional cut will show that when the piston is moving in one direction the ring valve on that face of the piston which is toward the direction of movement is closed, while that on the other face is open. This is exactly as it should be in order to discharge the compressed air from one end of the cylinder while taking in the free air at the other. The position of each valve is almost instantaneously reversed at the point when the stroke is reversed. This change in position takes place without springs or other influence than the *natural momentum of a piece of metal* which is carried in one direction and is instantly reversed. Place your pen-knife on the palm of your hand and thrust your hand forward and backward. You will thus have a simple illustration of the movement of this valve. You will see that at the instant that you reverse the direction of movement of your hand, the pen-knife continues on until the friction which holds it to your hand overcomes the momentum, which tends to continue its movement in the direction in which it was started.

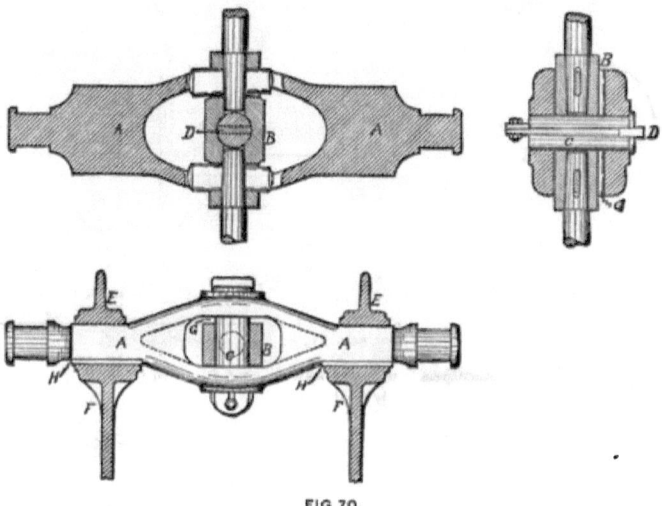

FIG 70.

Details of Crosshead,

Showing Independence of Swivel Block with Steam and Air Piston Rods from Crosshead.

This important improvement in air compressor crossheads maintains the weight of the crosshead on the guides notwithstanding natural wear, and prevents that weight from bearing on the rods.

Plan of Air Compressor, Air Receiver and Boiler.

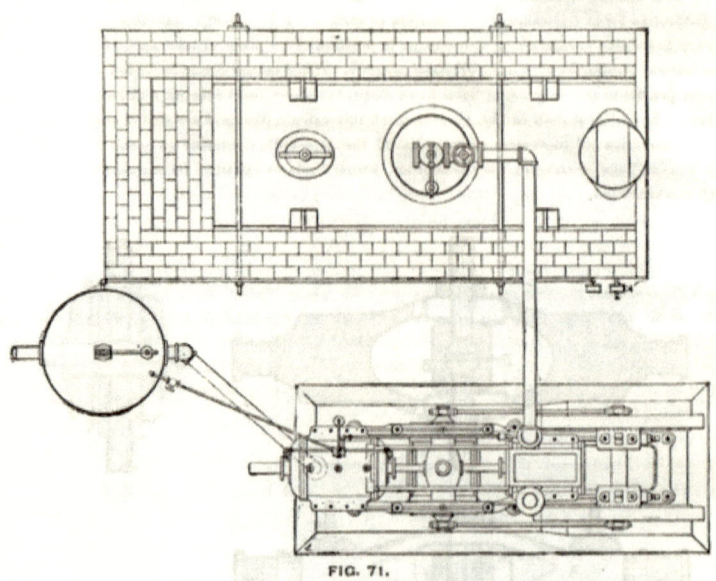

FIG. 71.

Showing Pipe Connections and Automatic and Adjustable Regulator for Air and Steam.

This Regulator takes the load from the engine when the air pressure reaches the desired point. Then a valve opens which permits air to pass from one end of cylinder to another and at the same time shut off the steam, so as only to permit enough steam to keep the compressor in motion; but as soon as the air pressure is lessened in the air receiver the air valve between ends of the cylinder closes, and at the same time the steam valve opens and compressor pumps in fresh air until the pressure again rises, thus securing a steady supply of air and high pressure.

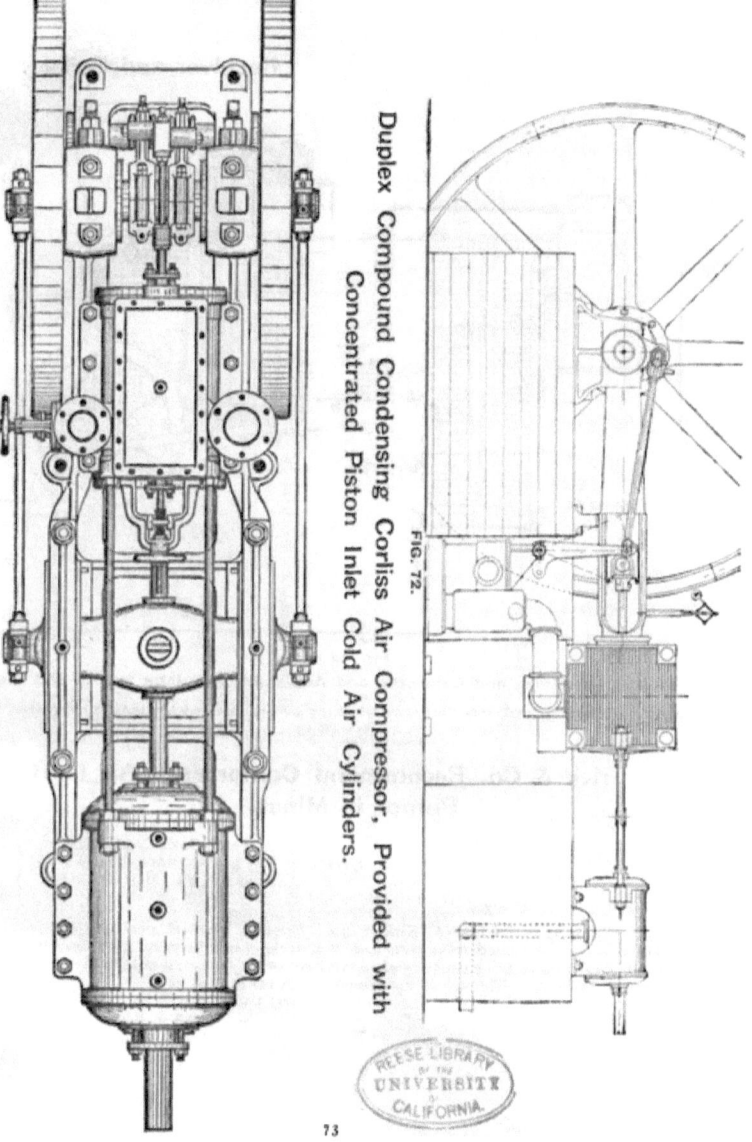

Duplex Compound Condensing Corliss Air Compressor, Provided with
Concentrated Piston Inlet Cold Air Cylinders.

FIG. 72.

Concentrated Piston Inlet Cold Air Compressor. Class "A," Steam Actuated.

FIG. 73.

Elevation of Air Compressor, Air Receiver and Boiler.

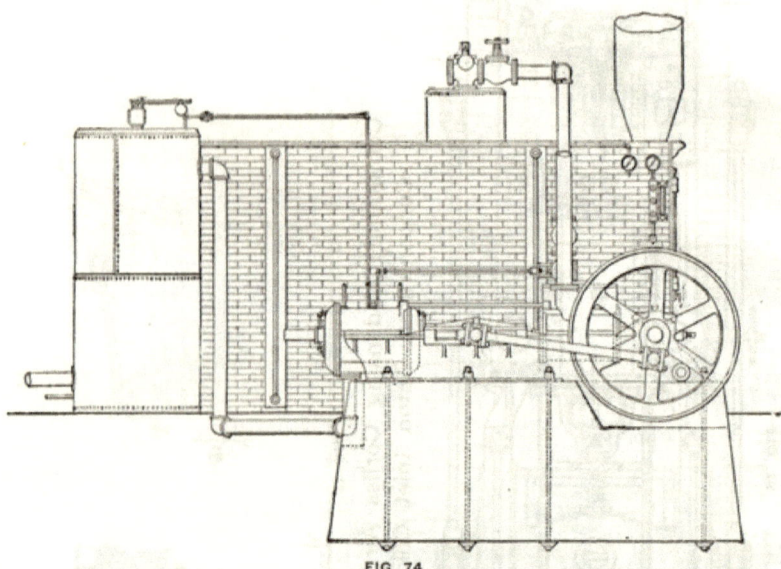

FIG. 74

Showing Pipe Connections, and Automatic and Adjustable Regulator for Air and Steam.

This regulator takes the load from the engine when the air pressure reaches the desired point.

H. C. Frick & Co. Recommend Compressed Air to Run Pumps in Mines.

H. C. Frick Coke Company,'
Office of Supt., Leisenring Mines No 1,
Leisenring, Pa., March 14, 1891.

Mr. Thomas Lynch, General Maurger :

Dear Sir:—Replying to letter of A. Kirk & Son, herewith attached, would say we have still the compressor at Leisenring purchased from them and it is working satisfactorily. We have three 10x12 Yough pumps on it running 10 to 15 strokes and carrying air to the fartherest one (nearly one mile), carrying fifteen pounds pressure. The size of the compressor is 10x30, No. 323.

Yours truly,

John A. Esser, *Superintendent.*

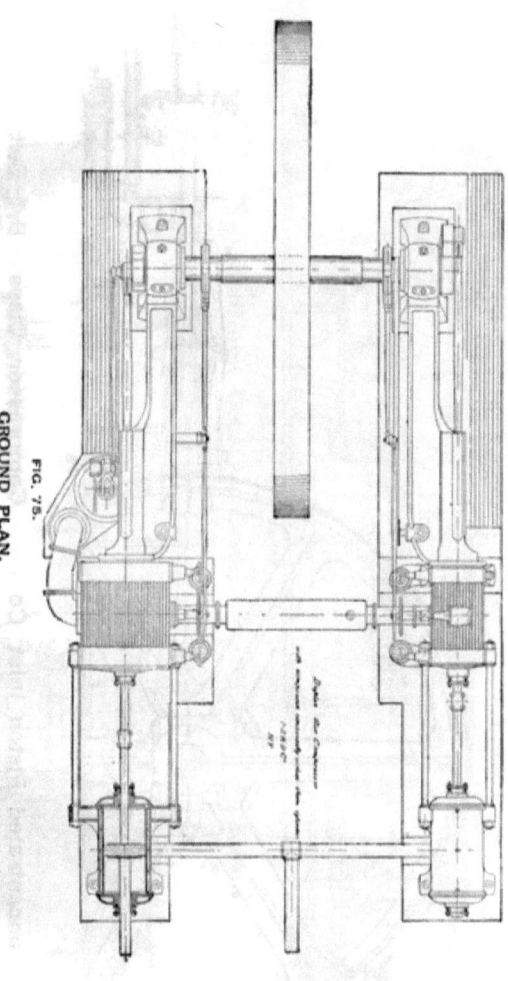

FIG. 75.

GROUND PLAN.

Duplex Compound Condensing Corliss Air Compressor, Provided with Concentrated Piston Inlet Cold Air Cylinders.

75

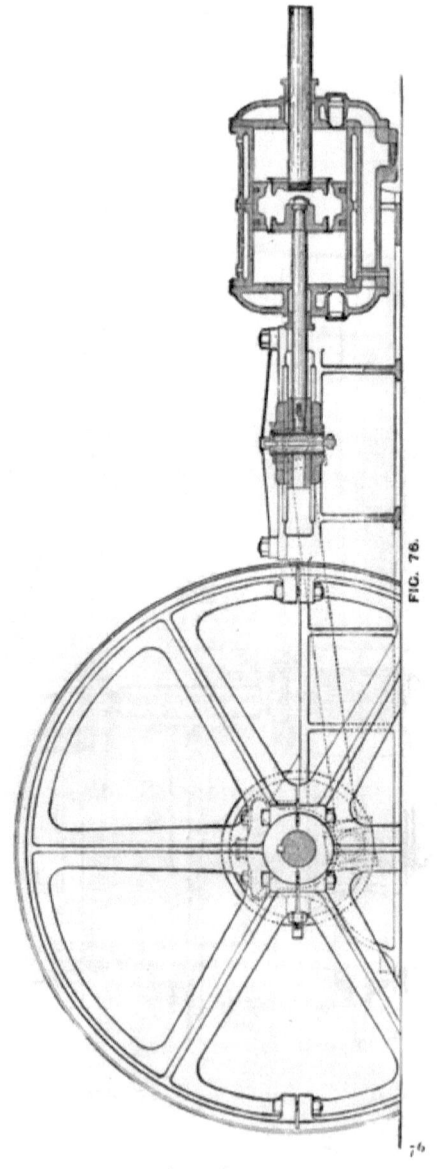

FIG. 76.

Concentrated Piston Inlet Cold Air Compressor. Class "B," Belt Actuated.

Driven by Water Power, Electricity or Stationary Engine.

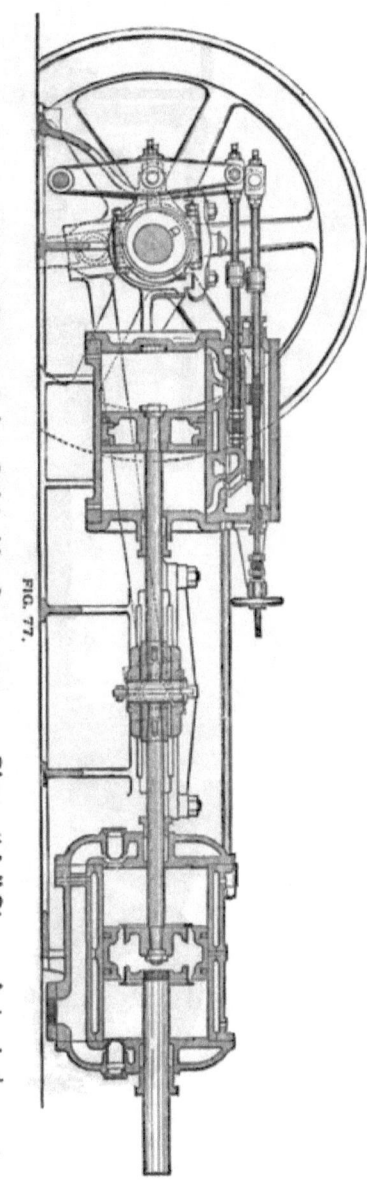

FIG. 77.

Concentrated Piston Inlet Cold Air Compressor. Class "A," Steam Actuated.

77

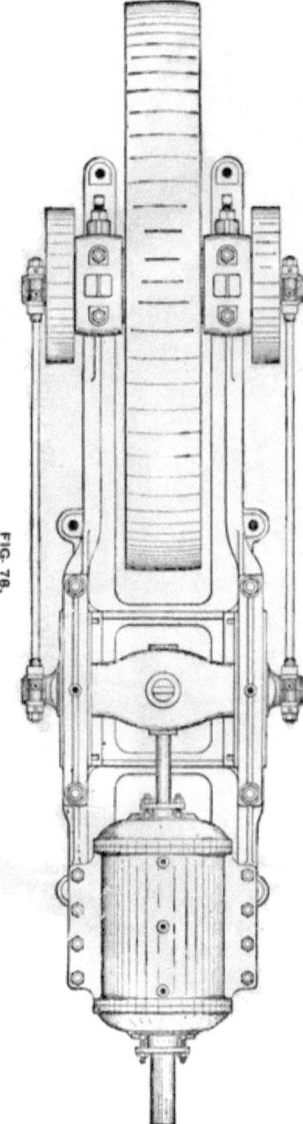

FIG. 78.

Concentrated Piston Inlet Cold-air Compressor. Class "B," Belt Actuated.

Driven by Water Power, Electricity or Stationary Engine.

FIG. 79.

Compound High Pressure Air Compressor.

For Air Pressures from 100 to 800 Pounds per Square Inch.

FIG. 80.

Air Compressor.

Enlarged Air Cylinders; Straight Line Type. Class "F." Made either with or without Water Jackets.

Used for Aerated Fuel Furnaces, Water Aeration, Pumping and Transferring of Acids by Air Pressure, Rock Drills and Pumps, and for running Sheep Shears, for which purpose large plants have been furnished in Australia.

FIG. 81.

Piston Inlet Cold Air Compressor.

Class "A," Steam Actuated. Showing Automatic Air Pressure Cut-off.

FIG. 82.

Air Cylinder of Piston Inlet Cold Air Compressor, Showing
Details of Automatic Pressure Regulator.

Class "A" or "B," Steam or Belt Actuated.

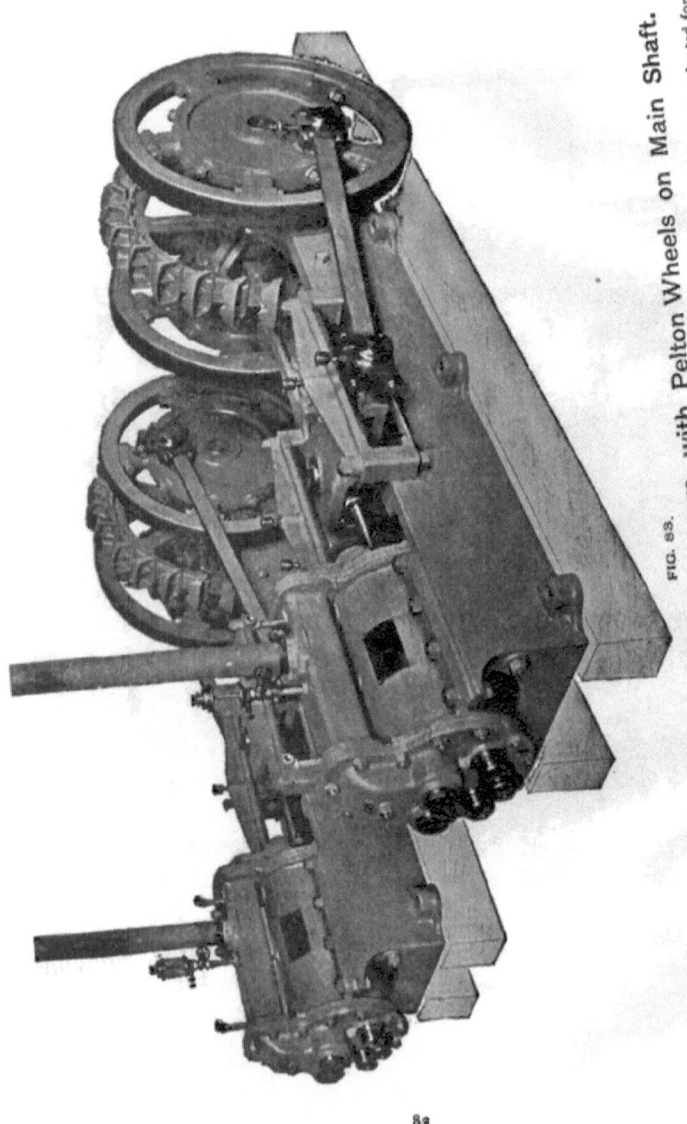

FIG. 83.

A Battery of Straight Line Air Compressors, with Pelton Wheels on Main Shaft. Particularly adapted for mines in inaccessible places, having mountain streams within reach. Air can be compressed by this machine at a place where plenty of water power can be had, and the air piped many miles to run machinery of any kind.

Sectionalized for mule transportation with single or double water nozzles for heads of fifty feet and upward.

FIG. 83½

A Mining or Tunneling Plant, Showing Boilers,

Compressor, Hoisting Engine, Air Receiver and Head Frame all Conveniently Arranged, so as to Economize Labor and have all under the personal supervision of one Chief Engineer.

83

OF POWER ROCK DRILLS.

All power drills (that is, drills driven by compressed air or steam) that I recommend, may be called striking drills and consist of a small, peculiarly-constructed steam cylinder, supported in slides which are mounted on three legs, A B C, fig. 84 (called the tripod). The cylinder is hung on feed-screw, D, and is

Drilling by Steam or Compressed Air.

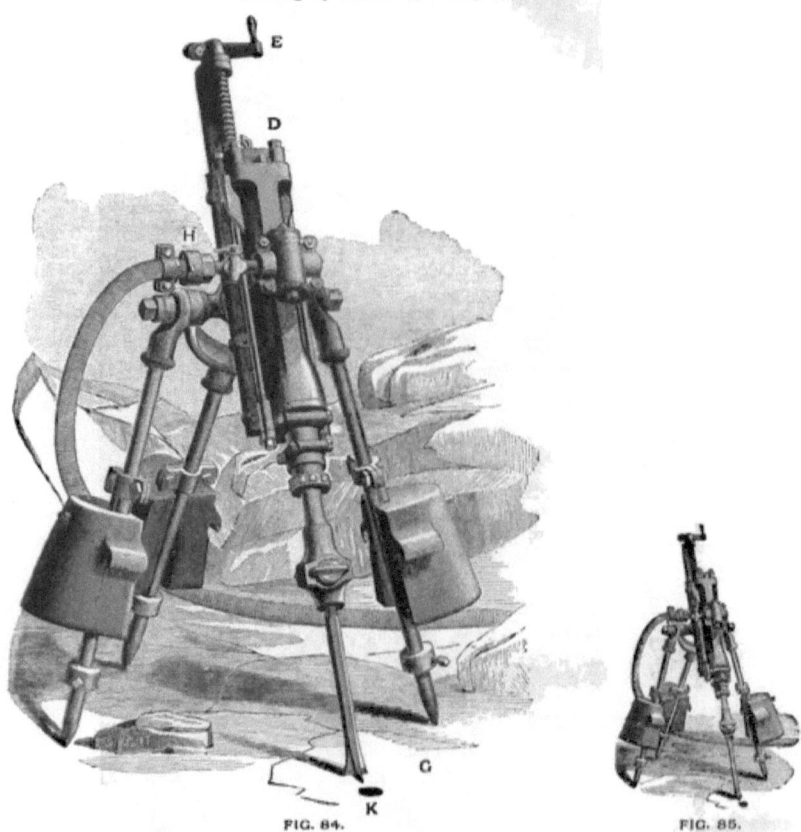

FIG. 84.   FIG. 85.

All Sizes of Rock Drills can be had from Fig. 84 to Fig. 85.

moved up or down. As the attendant turns feed-crank, E, the projecting end of piston is fitted with chuck to receive end of steel, G, and the piston is moved up and down from one hundred to three hundred revolutions a minute and the steel rotates once in every seven strokes. The duty of an attendant is mainly to start and stop the machine by turning the throttle valve, H, and to keep feed screw so fed that the piston head shall not strike head of cylinder, but feed it so that the full force of the blow shall be delivered on the rock, and as the rock is cut away and the hole K made, the crank F must be fed down just as the rock wears away. These drills are very well constructed, and with anything like proper care seldom get out of order. The running of one is very easily learned; any person of moderate mechanical skill can learn to run them in a few hours.

By placing column (shown in fig. 86), in center of heading and using long or short steels, and slacking the clamp bolts of the arm, the drill can be swung all around and made fast at any point of the circle, and then by slacking the saddle clamp bolts and raising or depressing point of drill nearly twenty holes can be made to good advantage at one setting of column.

FIG. 86.

Running a Heading in a Tunnel.

ARTHUR KIRK'S WATER INJECTOR FOR POWER ROCK DRILLS.

How to get sufficient water into the hole to wash out the dust made by a drill driven by steam or compressed air at from one hundred to three hundred strokes per minute, has long been an unanswered question. But the apparatus shown in fig. 88 solves this perplexing question completely.

DIRECTIONS FOR USING THE KIRK WATER INJECTOR.

Water being placed in bucket B, which is furnished with hand pump A. When the driller sees that water is needed, he pulls up the handle A two or three strokes, which fills the hose C and pipe D, and passes down through small brass pipe F into the hole and is discharged with great force at about ten inches from bottom of hole.

When the screw has run out and the steels must be changed, unscrew the union F and remove the short brass pipe from hole and place it in its proper place on rack-board M ; then change steels, putting in the next longer length steel, and then attach a corresponding length of pipe from rack M to union F,

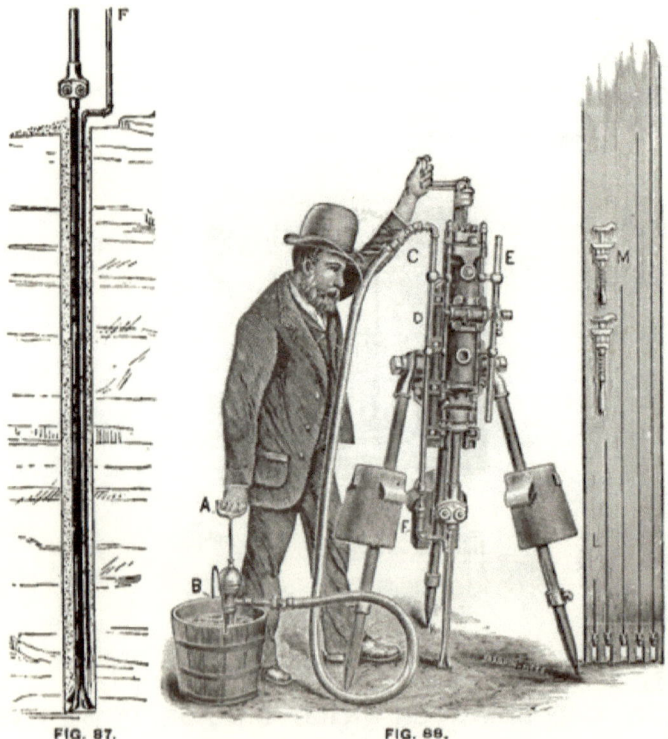

FIG. 87.   FIG. 88.

and so on until desired depth of hole is reached, then disconnect last brass pipe, place it in its proper place on rack, and thus every pipe is kept in place, protected from injury in the quarry and ready to be easily moved to another location, ready for immediate use.

All the clean water thus passes direct to bottom of hole without coming in contact with the dirty water coming out of hole, and thus completely prevents the forming of mud in the hole, and leaves the drill to dance free as in air, and strike the rock at bottom of hole with its full force, and thus by preventing the sticking of drill prevents all unnecessary strain on the rotation parts of drill, and the whole drill will last longer with less repairs, and make from ten to fifteen feet more hole per day than can be done without this arrangement.

A connection can also be made from the steam or air hose in advance of the throttle valve or cock to the pipe, C, by which the condensed steam can be drawn off and forced to bottom of hole, thus giving better steam, and also getting up a pressure in bottom of hole which will at once free the drill, should it stick, as it often does when the steel is dull and in need of sharpening.

An enlarged view of its action is shown in fig. 87, where F shows the water pipe drowning the dust at bottom of hole as soon as made.

I have lately witnessed a drill running in such a peculiar, tough, adhesive rock that it would have been utterly impossible to use a power drill on it without this plan of injecting the clear water to bottom of the hole. An effort was made to run the starter without working the water through the injector, but before it was half down it was working mud and running slow, and when the hole was finally worked down and the starter removed, it came out of the hole more like the plunger of a pump, with long ball of tough, adhesive mud sticking all over it, and the hole had to be swabbed out before another steel could be put in. After that the small brass pipes were regularly changed every time steel was changed, and the drill worked splendidly to full depth of hole without using a swab stick.

OIL AND TOOLS FOR POWER DRILLS.

Oil for steam or air drills should be of best No. 1 cylinder oil. Almost all cheap oils have acids in them which eat the cylinders, packing rings or valve seat, and thus cause the machine to leak steam and lose power, and need to be repaired long before they would if nothing but first-class oil had been used. Other cheap oils gum, and cause the drill to stick and lose time.

The oil can should be a thumb squirt can, and oil should be used sparingly and only on the parts exposed to friction. Oil should not be seen on any other part of the machine.

Every steam or air drill should be supplied with a thirteen and fifteen inch Beamer and Call combination pipe and monkey wrench, shown at M, fig. 88, page 86. They are the best I have seen used and also one-eight inch S monkey wrench.

WORK FOR WHICH DRILL IS BEST SUITED.

There are Drills of Eight Sizes, Adapted to all the Requirements of Rock Work.

Size "A," 1¼ inches diameter of cylinder. This is the smallest and lightest rock drill made. It is called "Baby." It is intended for plug and feather work in quarries and for block holing—that is, drilling holes in loose boulders or large pieces of broken rock, and in any kind of rock work where shallow holes are required. It should not be used to drill holes deeper than three or four feet, or of larger diameter than 1¾ inches. This drill has been used to put in holes ten feet deep, but it is out of place in such work, as a larger drill would do more work in the same time and would last longer. It is light in weight—easily handled by one man. It is mounted either on tripod column, shaft-bar or quarry-bar, as desired. It is sometimes used for trimming the walls, roofs or floors of mines or tunnels.

Size " B," 2½ inches diameter of cylinder. This drill is intended for work where holes are put in to a depth of from four to six feet, and from 1 to 1¾ inches in diameter. It is used in mining narrow veins of ore, in driving tunnels of small section ; in quarries, for drilling plug and feather holes; also, for blast holes. In mines it is principally used on shaft-bar or single-screw column. In quarries it is used on quarry-bar for drilling a line of holes.

Size " C," 2¾ inches diameter of cylinder. This drill is used for holes from six to ten feet in depth and from one to two inches in diameter, being one size larger than " B." It is used in small tunnels and narrow veins, where the rock is very hard. It is largely used in quarries, mounted on tripod, quarry-bar and gadding frame. It is the proper size for standard gadder used in marble quarries. It is more largely used on the quarry-bar than any other size.

Size " D," 3 inches diameter of cylinder. This is the average size, more largely used than any other. It is especially adapted for general mining work. It is used in driving the heading of tunnels where the rock is not so hard as to call for a larger drill. It is largely used in shaft sinking and in driving drifts of six or eight feet diameter. It is intended for drilling holes to a depth of from 10 to 12 feet, and from 1¾ to 2 inches in diameter. It is used in quarries, mounted either on tripod or quarry-bar.

Size " E," 3¼ inches diameter of cylinder. This drill being a little larger than the " D," is intended to substitute the " D " in work where a little larger drill is required, because of hard rock or deeper holes. It is used to put in holes averaging about 12 feet in depth and from 1¾ to 2½ inches in diameter

Size " F," 3½ inches diameter of cylinder. This drill is intended for holes averaging about 16 feet in depth and from 1½ to 3 inches in diameter. It is the preferred size for large open-cut excavations in railroad building, canals, etc. It was the adopted size for driving the headings on the New York aqueduct tunnel. It is the best size for all tunnels of not less than 10 feet diameter and in hard rock. It is the preferred size in granite quarries, and is frequently mounted on the quarry-bar for broach channeling. It is the size used on our bar channeler There are two patterns of this size : one the " F " and the other the " F 2." The latter is a special drill, made heavier and for heavier work.

Size " G," 4¼ inches diameter of cylinder. This drill is made with the automatic feed attachment, except when otherwise specified. It is intended for holes from 20 to 30 feet in depth and from 2 to 4 inches in diameter. It is used almost exclusively on tripod for surface work in extremely hard rock. It was preferred by the contractors in the excavations on the Pennsylvania Railroad at Bergen Cut, and is the best size for all similar work—that is, where the cutting is heavy and the rock very hard. It is largely used in granite quarries for the deep-hole work. It is preferable to the 3½ inch " F " for holes even fifteen feet in depth, where the rock is as hard as trap or granite and the work extensive.

Size " H," 5 inches diameter of cylinder. This is the largest rock drill made. Its special application is in submarine work. It is sometimes used in very heavy railway or canal rock cuts. It is intended for holes from 25 to 50 feet in depth and from 3 to 6 inches in diameter. It is made with the automatic feed attachment, except when otherwise specified.

FIG. 89.

SINGLE QUARRY BAR.
The Lightest, Simplest—Very Strong and Portable.

FIG. 90.

Bar Channeler Making a Horizontal Cut.

FIG. 91.

Running a Tunnel with Power Drills and Compressed Air.

FIG. 92
Sinking a Large Shaft with Four Power Drills Run by Compressed Air.

Showing the different positions drills can be used in. The drill standing on left hand of cut is standing on only the two front legs, the third or hind leg turned up along face of rock out of the way, so as to permit the drill to make a hole close in to face of rock. The drill on right hand is drilling a perpendicular hole. The center drill is set to drill a slanting hole at bottom of rock, and the forth drill is mounted on a horizontal bar.

FIG. 93.

Quarrying with the Bar Channeler.

The above cut illustrates a typical case of quarrying by the channeling process.   In this case it is especially noticeable that the channeling machine which has made such regular benches in the quarry is a light and inexpensive Bar Channeler.

The advantages of a quarry in such regular shape as this will be apparent to any experienced quarry-man.   The stone is removed in blocks ready for the market.   A uniform system is followed in removing the stone from its place in the bed, enabling the foreman to calculate with accuracy how much stone will be shipped within a definite length of time.   Besides these advantages, there is a great saving in expense.   The stone costs less per cubic foot when brought out in blocks, and there is little or no waste to be thrown into the dump.   A quarry which is run by the channeling process requires, besides the Channeling Machine, a Rock Drill mounted on either a Tripod, a Gadder Frame or Quarry Bar, as the stone may require.   This is all the machinery necessary for the removal of the stone from its place in the bed.

FIG. 93½.

Sinking a shaft with power drills and showing the wonderful adaptability of a tripod, the drill working so close to face of rock that there is only space for one leg of tripod parallel to drill steel. One leg is braced back against the rock and the other leg extending horizontally and braced against the rock.

OFFICE OF SCHWEYER & LIESS,
PRODUCERS OF PENNSYLVANIA BLUE MARBLE,
KING OF PRUSSIA, MONT. CO., PA., April 9, 1890.

GENTLEMEN:—In our quarry we removed 3,704 cubic yards of marble in one year with one of your screw frame channelers, one bar channeler and one gadder. This required about 10,000 square feet of channeling or we removed about ten cubic feet of marble per square foot channeled. The average cost per square foot of channeling is thirty-seven cents. The average cost per cubic foot of marble on the quarry bank is about ten cents. Before we used your machinery in our quarry, the average cost of our marble on the bank was twenty-five cents. We now are able to remove four times as much marble in the same time that we could by quarrying by hand. The cost per square foot channeled appears high, because we only use a thirty horse power boiler for running our three machines, and so far had a very uneven floor in our quarry to work the machines on. We think that when we get a level floor in our quarry (which we are working for), and increase our power, we shall be able to channel almost double the amount of square feet in the same time. If so, it will very much reduce the average cost per square foot channeled, and will increase the difference between quarrying by hand and quarrying by machines in proportion.　　　　Yours respectfully,　　　　SCHWEYER & LIESS.

H. I. Schweyer, *Manager.*

IMPORTANT FEATURES OF THE AUXILIARY VALVE DRILL.

The auxiliary valve drill is, strictly speaking, a drill for hard rock. The design and purpose of the inventor was to strike a hard blow, and to so build the machine that it would stand hard usage for years. The experience of the last four years has conclusively proved that not only is the drill remarkably efficient in cutting capacity, but that it does its work after years of use equally as well as when new.

The auxiliary valve is the only rock drill in the world which combines the independent valve operated through an auxiliary valve, and which contains a release rotation. These two features are the most important as distinguishing it from other rock drills.

VALUABLE POINT IN THE AUXILIARY VALVE.

1. The auxiliary valve strikes an uncushioned blow. The valve is held in such a position that while the piston carrying the cutting tool is moved toward the rock the exhaust remains open at one end, while the full pressure acts on the other end until the blow is struck, at which time the valve immediately reverses. There is no such thing in the drill as striking a blow upon a cushion of steam or air in front end of cylinder. It must hit the rock and does it before the steam or air enters the front end. It does not use steam or air expansively, but has the benefit of full pressure to strike the blow and to recover from broken or crooked holes.

2. It has an auxiliary valve operated by shoulders upon the piston. The auxiliary valve and its valve seat are entirely independent of the main valve and seat. The auxiliary is the trigger to the main valve. It opens or closes the steam or air passages, releasing the pressure from one end or the other of the main valve. The pressure bears it upon its seat; hence its wear is uniform and cannot produce leakage. The auxiliary valve being light, of steel, and moving on the arc of a circle through contact with the piston operating tangentially, it is easily moved, does not wear rapidly, and never breaks. It is inexpensive and readily duplicated.

3. A perfect valve motion. Using a round piston made of steel and hardened, fitting plug like in the ends, a large opening is effected by a slight movement of the valve. Being perfectly balanced, there is little or no wear. There is not a single instance on record of an auxiliary valve breaking.

4. A short or long stroke can be obtained at will by turning the crank and feeding the cylinder toward the rock. This is a most important feature. A short stroke is of great advantage in starting or blocking out holes.

5. A new rotating devise with a release movement, which prevents twisting of the spiral bar or breaking of pawls and ratchets. When a rock drill strikes a hard blow upon an uneven surface there is a tendency sometimes to twist the steel in the opposite direction to that in which it rotates. The effect of such a blow on this drill is simply to turn the back head around, overcoming the friction of the back head springs, when with a rigid rotation it might twist the rifle bar or break the pawls and ratchets.

6. Two strong steel springs are used in place of buffers. These springs are placed on the back head and are connected with the front head through the side bolts; hence a blow upon either the front or the back head is cushioned by the springs, thus preventing breakages.

7. Its construction is such that it can be taken apart and put together in a few minutes, the parts eing few and simple. It is not liable to get out of order. It is easily handled and understood by the runner.

FIG. 94.

"BABY" DRILL.

Size "A," 1¼ inch Cylinder. The simplest and lightest Drill made.
For Plug and Feather Holes, and other light work.

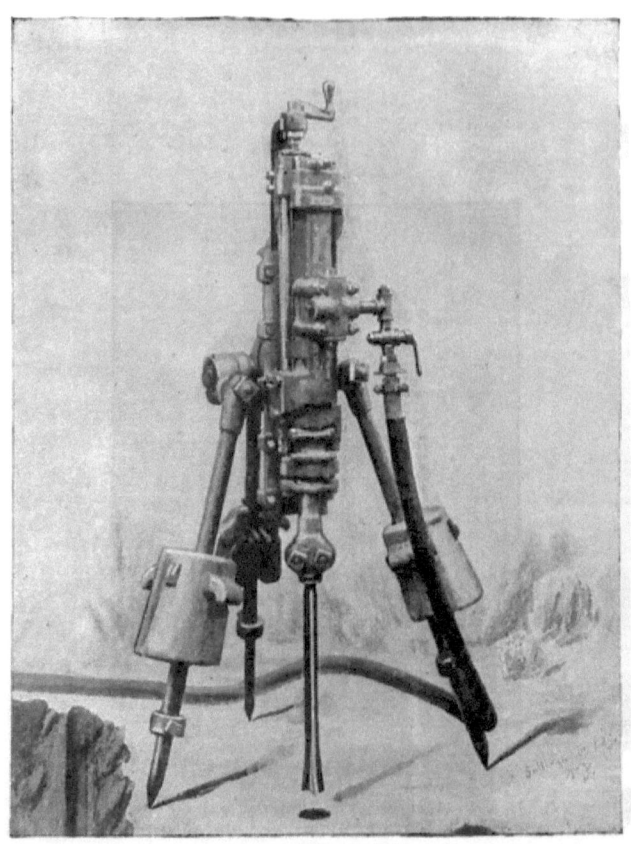

FIG. 95.

A. D. AUXILIARY VALVE DRILL.

Ready to start, showing Hose attached and Starter Bitt in, and has just made a few inches of a hole. Can be run to strike 200 blows per minute.

FIG. 96.

Starting a Quarry with Power Drill Machinery.

That useful quarrying machine, the Bar Channeler, is shown in comparison with hand trenching.

FIG. 97.

Flat Hole Work with a Power Drill.

FIG. 98.

Breaking up Large Blocks of Granite

Into sizes ready for the market. Showing Power Drill and Bar at work at the quarries c.
Brandywine Granite Company, Wilmington, Del.

MORAN FLEXIBLE STEAM JOINT.

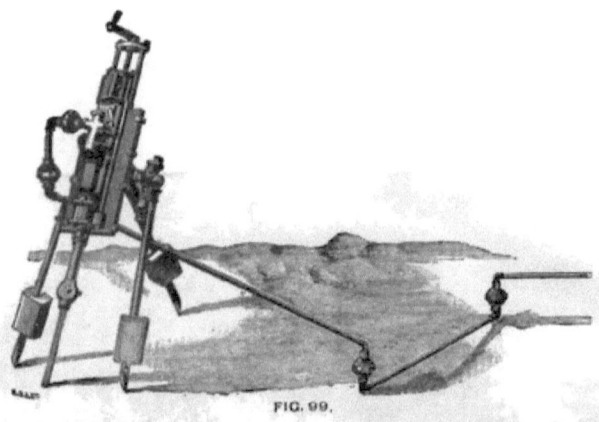

FIG. 99.

It is claimed that the Moran Flexible Steam Joint is reasonable in original cost; economical in steam saving and actual wear; more efficient than any rubber connection, because of its simplicity and indestructibility, and saving of care, worriment and labor proportionately; and with unquestionable better results, because it will admit of any desired pressure.

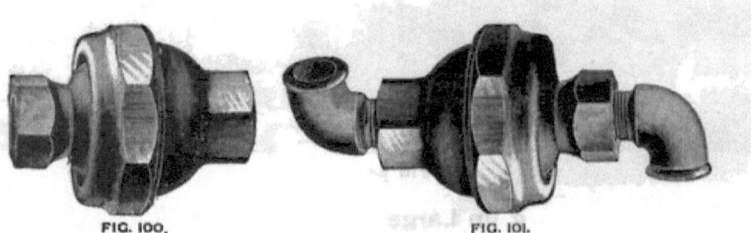

FIG. 100.

Perspective View of Steam Joint.

FIG. 101.

Sectional View of Same.

The Moran Joint was invented to cover the many points where the flexible joint is a necessity.

Attention is called to a few of the many uses wherein it might be called a necessity.

Steam drills—Making a flexible and durable line of pipe, and doing away with the present unsubstantial and expensive rubber hose system.

Expansion steam joint—For long lines of steam pipe, two of these joints will allow any distance of expansion and contraction desired.

ROCK DRILL HOSE AND COUPLINGS.

For conveying steam or air to drills.   One length required for each machine.

	A	B	C	D	E	F	G	H
Letter indicating size of drill......................	A	B	C	D	E	F	G	H
Diameter of cylinder......................Inches.	1¾	2½	2¾	3	3⅛&3¼	3½	4¼	5
Size of hose...................................	¾	1	1	1	1	1	1	1
FOR STEAM, 50 feet, 5-ply special steam hose (marline wound or canvas covered)	$30 00	$30 00	$30 00	$33 00	$33 00	$33 00	$33 00	$33 00
Same with patent lightning couplings attached	33 50	33 50	33 50	37 00	37 00	37 00	37 00	37 00
FOR AIR, 50 feet special air hose (marline wound or canvas covered)	20 00	20 00	20 00	25 00	25 00	25 00	25 00	25 00
Same with lightning couplings attached...............	23 50	23 50	23 50	29 00	29 00	29 00	29 00	29 00

STEAM HOSE.

Prices subject to market changes.   Winding with marline, 10 per cent. additional.

Size, inside diameter......................Inches.	½	¾	1	1¼	1½	1¾	2	2½
FOUR-PLY, extra heavy. Price per foot............	$0 33	$0 40	$0 53	$0 66	$0 79	$0 92	$1 05	$1 31
FIVE-PLY, special. " " ".............	0 41	0 55	0 66	0 83	0 99	1 14	1 31	1 64
SIX-PLY, extra strong. " " ".............	0 49	0 65	0 80	0 99	1 18	1 37	1 57	1 95

FOUR-PLY SPECIAL AIR HOSE. (Engine Hose.)

Winding with marline or canvas, 10 per cent. additional.

Size, inside diameter.........Inches.	½	¾	1	1¼	1½	1¾	2	2¼	2½	2¾	3	4
Price, per foot.......................	$0 20	$0 25	$0 34	$0 42	$0 50	$0 58	$0 67	$0 71	$0 84	$0 92	$1 00	$1 34

EXHAUST HOSE, THREE-PLY.

Size, inside diameter.....Inches.	½	¾	1	1¼	1½	1¾	2	2¼	2½	2¾	3	3½	4
Price, per foot..............	$0 17	$0 20	$0 27	$0 34	$0 40	$0 48	$0 54	$0 60	$0 67	$0 74	$0 80	$0 94	$0 97

All of the above hose in stock in lengths of 25 and 50 feet.   Other lengths made to order.

EUREKA STEAM HOSE.

Size	½	¾	1	1¼	1½	1¾	2	2¼	2½
Price...	$0 40	$0 48	$0 60	$0 72	$0 88	$1 04	$1 20	$1 40	$1 60

This hose has a specially prepared rubber tube covered with a multiple woven cotton seamless tube, which does not weaken by the wear upon the outside.

SPIRAL WOUND WIRE COVERED STEAM HOSE.

Size	¾	1	1¼	1½	2	3
Price, per foot, three-ply...	$0 44	$0 60	$0 76	$0 90	$1 20	$1 80
" " " four-ply..	0 60	0 75	0 88	1 00	1 50	2 40

PRICE OF ROCK DRILL

With Complete Outfit for Surface Work Where Holes are Put in About Twelve Feet Deep.

Price.

One standard size "E" rock drill (diameter of cylinder 3½ inches), mounted on universal joint tripod, furnished complete with valves, weights and wrenches.. $350 00

One set of drill steels (six pieces), fitted with bits and shanks, for drilling holes up to twelve feet in depth. 25 00

Fifty feet of one-inch rock drill steam hose, made either with friction back or marline wound as preferred, furnished with patent security couplings attached. 37 00

One set of blacksmith's tools for sharpening drill bits, one sand pump, band for entering piston, and extra parts.. 15 00

Shipping Weight, lbs.

Total.. $427 00   1113

One ten horse-power vertical tubular boiler, furnished complete with stack, grates, guages and injector attached, boiler being supplied with all fittings, ready to fire 222 00   1900

SIXTEEN FEET DEEP.

One standard size "F" rock drill (diameter of cylinder 3½ inches), mounted on universal joint tripod, furnished complete with valves, weights and wrenches.. 375 00

One set of drill steels (eight pieces), fitted with bits and shanks, for drilling holes up to sixteen feet in depth.. 48 00

Fifty feet of one-inch rock drill steam hose, made either with friction back or marline wound as preferred, furnished with patent security couplings attached 37 00

One set of blacksmith's tools for sharpening drill bits, one sand pump, band for entering piston, and extra parts.. 15 00

Total........ .... $475 00   1307

One fifteen horse-power vertical tubular boiler, furnished complete with stack, grates, guages and injector attached, boiler being supplied with all fittings, ready to fire 276 00   2800

TWENTY-FIVE FEET DEEP.

One standard size "G" rock drill (diameter of cylinder 4¼ inches), mounted on universal joint tripod, furnished complete with valves, weights and wrenches. 430 00

One set of drill steels (ten pieces), fitted with bits and shanks for drilling holes up to twenty-four and one-half feet in depth.............................. 112 00

Fifty feet of one-inch rock drill steam hose, made either with friction back or marline wound as desired, furnished with patent security couplings attached. 37 00

One set of blacksmith's tools for sharpening drill bits, one sand pump, band for entering piston, and extra parts.. 15 00

Total $594 00   2190

One fifteen horse-power vertical tubular boiler, furnished complete with stack, grates, guages and injector attached, boiler being supplied with all fittings, ready to fire 276 00   2700

PRICE OF COMPLETE PLANT OF MINING MACHINERY

For Operating Six Size "D" (Three-Inch Cylinder) Rock Drills by Compressed Air.

DRILLS, ETC.

	Price.	Shipping Weight, lbs.
Six standard mining drills, size "D" (three-inch cylinder), complete with valves and wrenches (unmounted), $275 each......................	$1650 00	
Six new style, double screw tunnel columns, complete with arms and clamps, $60 each	360 00	
Price of shaft bars if required, $50 each.		
Price of tripods with weights, $50 each.		
Six sets of fitted drill steels for drilling holes up to ten feet in depth, $16 per set	96 00	
Six lengths (50 feet each) one-inch special air hose, with couplings attached, $29 per length..................................	174 00	
One set of blacksmith's tools for sharpening drill bits...................	10 00	
Extra duplicate parts..	50 00	
Total for drill outfit....................................	**$2340 00**	**4287**

AIR COMPRESSOR, ETC.

One standard class "A" straight line air compressor, size P, cylinders sixteen-inch diameter, stroke eighteen inches, of capacity to run seven three-inch cylinder "D" drills, as specified in catalogue...	2500 00	
One steel air receiver, diameter forty-two inches, length ten feet; furnished complete with guages, safety valve and fittings.............	202 00	
One seventy horse-power horizontal tubular boiler, of half-front pattern, furnished complete with stack, grates, guages, rollers, brackets, guy rods and all fittings, including injector attached, being complete, ready to fire...	1076 00	
Estimated cost of pipe and fittings to connect boiler with air compressor and compressor with receiver..................................	100 00	35000
Total for compressor outfit.............................	3878 00	39287
Cost of complete plant.................................	$6218 00	
Five hundred feet of four-inch air pipe to extend from air receiver into the mine (price subject to market change), 42 cents per foot..........	210 00	
Valves and fittings for pipe line.....	75 00 285 00	5500
Total	$6503 00	44787

SPECIFICATIONS AND PRICES OF DRILL STEELS.

SIZES AND PRICES.

For Drill A.} 1¾". 2".

⅝" x 3" Shank. Feed, 12".

LENGTH OF STEEL.	SIZE OF STEEL.	WEIGHT	PRICE. SINGLE.	PRICE. SET.
1 ft.	¾ in.	3 lbs.	$1 43	
2 "	¾ "	5 "	1 65	$3 08
3 "	¾ "	7 "	1 87	4 95

For Drill B. 2¼".

⅞" x 4¾" Shank. Feed, 20".

1' 8"	1 in.	7 lbs.	$1 87	
3' 4"	⅞ "	9 "	2 09	$3 96
5'	⅞ "	13 "	2 53	6 49

For Drill C. 2¾". 2¾".

1" x 5" Shank. Feed, 20", 24".

1' 8"	1⅛ in.	9 lbs.	$2 09	
3' 4"	1⅛ "	15 "	2 75	$4 84
5' 0"	1 "	16 "	2 86	7 70
6' 8"	1 "	21 "	3 41	11 11
8' 4"	1 "	26 "	3 96	15 07

For Drill D. 3".

1" x 5" Shank. Feed, 24".

2'	1⅛ in.	10 lbs.	$2 20	
4'	1⅛ "	18 "	3 08	$5 28
6'	1 "	19 "	3 19	8 47
8'	1 "	24 "	3 74	12 21
10'	1 "	30 "	4 40	16 61
12'	1 "	35 "	4 95	21 56

For Drill F. 3¼".
For Drill D.} 3".
For Drill E.} 3¼".

1⅛" x 5" Shank. Feed, 24".

LENGTH OF STEEL.	SIZE OF STEEL.	WEIGHT	PRICE. SINGLE.	PRICE. SET.
2 ft.	1¼ in.	11 lbs.	$2 31	
4 "	1¼ "	20 "	3 30	$5 61
6 "	1⅛ "	24 "	3 74	9 35
8 "	1⅛ "	31 "	4 51	13 86
10 "	1⅛ "	38 "	5 28	19 14
12 "	1⅛ "	46 "	6 16	25 30
14 "	1⅛ "	54 "	7 04	32 34

For Drill F.} 3¼". 3½".

1¼" x 5¾" Shank, } Feed,
1¼" x 5" Shank, } 24".

2 ft.	1⅜ in.	14 lbs.	$2 64	
4 "	1⅜ "	25 "	3 85	$9 49
6 "	1⅜ "	36 "	5 06	11 55
8 "	1¼ "	39 "	5 39	16 94
10 "	1¼ "	48 "	6 38	23 32
12 "	1¼ "	57 "	7 37	30 69
14 "	1¼ "	66 "	8 36	39 05
16 "	1¼ "	75 "	9 35	48 40
18 "	1¼ "	84 "	10 34	58 74
20 "	1¼ "	93 "	11 33	70 07

For Drill G. 4¼".

1½" x 6" Shank. Feed, 30".

LENGTH OF STEEL.	SIZE OF STEEL.	WEIGHT.	PRICE. SINGLE.	PRICE. SET.
2 ft.	1⅝ in.	22 lbs.	$3 52	
4½ "	1⅝ "	39 "	5 39	$8 91
7 "	1⅝ "	59 "	7 59	16 50
9½ "	1½ "	65 "	8 25	24 75
12 "	1½ "	81 "	10 01	34 76
14½ "	1½ "	98 "	11 88	46 64
17 "	1½ "	114 "	13 64	60 28
19½ "	1½ "	131 "	15 51	75 79
22 "	1½ "	148 "	17 38	93 17
24½ "	1½ "	165 "	19 25	112 42
27 "	1½ "	182 "	21 12	133 54

For Drill H. 5".

1½" x 6" Shank. Feed, 30".

2 ft.	1⅝ in.	22 lbs.	$3 52	
4½ "	1⅝ "	39 "	5 39	$8 91
7 "	1⅝ "	59 "	7 59	16 50
9½ "	1½ "	65 "	8 25	24 75
12 "	1½ "	81 "	10 01	34 76
14½ "	1½ "	98 "	11 88	46 64
17 "	1½ "	114 "	13 64	60 28
19½ "	1½ "	131 "	15 51	75 79
22 "	1½ "	148 "	17 38	93 17
24½ "	1½ "	165 "	19 25	112 42
27 "	1½ "	182 "	21 12	133 54
29½ "	1½ "	200 "	23 10	156 64
32 "	1½ "	217 "	24 97	181 61

DESCRIPTIVE TABLE OF ROCK DRILLS.

Letters indicating size	A	B	C	D	E	F	F2	G	H
Diameter of cylinder	1¾ in.	2½ in.	2¾ in.	3 in.	3⅛ in.	3½ in.	3½ in.	4¼ in.	5 in.
Length of stroke	4 "	5 "	5 "	6 "	6 "	6½ "	6½ "	8 "	3 "
Extreme length of drill from end of crank to end of piston	36 "	42 "	43 "	48 "	48 "	53 "	53 "	60 "	60 "
Diameter of supply inlet	½ "	¾ "	¾ "	1 "	1 "	1 "	1 "	1 "	1 "
Weight of machine	107 lbs.	175 lbs.	227 lbs.	272 lbs.	273 lbs.	372 lbs.	420 lbs.	620 lbs.	693 lbs.
Weight of tripod, without weights	65 "	155 "	155 "	155 "	198 "	198 "	245 "	260 "	260 "
Shipping weight of drill, tripod and weights, complete	304 "	645 "	697 "	742 "	786 "	900 "	1019 "	1234 "	1493 "
Approximate strokes per minute, with 60 lbs. pressure at drill	400	360	325	325	325	300	300 "	250	250
Approximate weight of blow delivered on the rock at each stroke	200 lbs.	350 lbs.	500 lbs.	550 lbs.	625 lbs.	750 lbs.	750 "	1000 lbs.	1500 lbs.
Depth drilled without changing bits	12 in.	20 in.	20 in.	24 in.	24 in.	24 in.	24 in.	30 in.	30 in.
Average work done per 10 hours in granite, down holes, including time lost in setting drill and changing bits		50 feet	60 feet	60 feet	70 feet	70 feet	70 feet	70 feet	70 feet
Depth of vertical hole each machine will drill easily, from 1 to	3 feet	5 "	8 "	12 "	14 "	20 "	20 "	27 "	32 "
Diameter of holes drilled as desired. From	⅝ to 1¼ in.	1 to 1½ in.	1¼ to 2 in.	1¼ to 2 in.	1¼ to 2½ in	1½ to 3 in.	1½ to 3 in.	2 to 4 in.	3 to 6 in.
Diameter of drill steel used. Inch	⅝ & ¾ "	1 & ⅞ "	1⅛ & 1 "	1⅛ & 1 "	1¼ & 1⅛ in	1⅜ & 1¼ in	1⅜ & 1¼ in	1⅜ & 1½ in	1⅜ & 1½ in
Size of shanks	⅞x3 "	⅞x4¾ "	1x5 "	1x5 "	1⅛x5 in	1¼x5¾ in.	1¼x5¾ in.	1½x6 in.	1½x6 in.
No. of pieces in set of steels to drill holes of depths above stated	3	3	5	6	7	10	10	11	13
Approximate weight of one set steels to drill vertical holes of depths above stated	15 lbs.	29 lbs.	87 lbs.	136 lbs.	224 lbs.	537 lbs.	537 lbs.	1104 lbs.	1521 lbs.
Best size of boiler to give plenty of steam at high pressure	2 to 6 h. p.	8 h. p.	8 h. p.	10 h. p.	10 h. p.	10 to 15 h. p.	10 to 15 h. p.	15 h. p.	15 h. p.
Best size of supply pipe, carrying steam 100 to 200 feet	¾ to 1 in.	1 in.	1 in.	1 in.	1 in.	1 in.	1 in.	1½ in.	1½ in.
Price of drill, unmounted	$145	$225	$250	$275	$300	$325	$325	$375	$425
Price of adjustable tripod, complete	30	50	50	50	50	50	50	55	55
Price of drill and tripod, complete	175	275	300	325	350	375	375	430	480
Price of tunnel column (any length to 8 ft.) with arm and clamp	50	60	60	60	60	90	90	90	90
Price of plain shaft bar (to 9 feet) with clamp	50	50	50	59	50	84	84	84	84

SIZES AND WEIGHT OF COLUMNS.

Improved tunnel column, 4½ inches in diameter, 6 to 8 feet long, for mounting 2¾ to 3¼-inch drills, inclusive........ 6 feet long, 180 lbs.; 8 feet long, 200 lbs.

Four and one-half inch clamp, 40 lbs.; 4½-inch adjustable arm, for use in tunnel or drift, 60 lbs.

Four and one-half inch column, arm and clamp, complete	6 " 280 "	8 " 300 "	$ 60 00
Improved tunnel column, 5½ inches in diameter, 6 to 8 feet long, for mounting 4¼ and 5-inch drills, inclusive	6 " 240 "	8 " 260 "	
Five and one-half inch tunnel clamp, 70 lbs.; 5½-inch adjustable arm, for use in tunnel or drift, 90 lbs.			
Five and one-half inch column, arm and clamp, complete	6 " 400 "	8 " 420 "	90 00
Plain shaft bar, 5½ inches in diameter, 6 to 8 feet long, for mounting 4¼ and 5-inch drills, inclusive	6 " 215 "	8 " 235 "	84 00
Plain shaft bar, 4½ inches in diameter, 6 to 8 feet long, for mounting 2¾ to 3¼-inch drills, inclusive	6 " 160 "	8 " 180 "	50 00
Three-inch column, arm and clamp, for mounting 2¾ and 2½-inch drills, inclusive		200 "	50 00
Quarry bar for drill "A" to "C," for gadding, channeling and plug and feather work on dimension stone		500 "	175 00
Standard quarry bar for drills "C" to "G," inclusive			250 00

FIG. 102.

DIRECTIONS FOR OPERATING
ROCK DRILLS.

A successful driller must know how to *feed*, *place* and *care* for his machine properly.

First, adjust the tripod or column in place *firmly* and make sure that all bolts are thoroughly tightened before attempting to start the drill; then connect the hose by screwing the coupling on to plug cock or valve, between which and the steam chest there should be a Tee placed in this position ⊣ (see fig. 95), the lower opening taking the valve, the side opening fitting into steam inlet by means of a short nipple; this leaves the top opening free for the introduction of oil to be blown into the steam chest. This opening must always be plugged, except when pouring in the oil.

Wipe the shank of the steel perfectly clean before inserting it into the chuck; force it in completely, and tighten the chuck bolt nuts thoroughly and EVENLY, but avoid wrenching them off. After packing the front head, screw the stuffing-box up just tight enough to prevent the escape of steam. Keep the piston-rod well oiled, also the interior of the drill through the thumb screw hole in the back head and the valve through the supply port. Use good oil which will not gum, keep the valve clean and free from grit. Before attaching the hose to the drill let the steam or air blow through to remove the water and dirt in the pipe. A valve should be attached to the iron pipe where it joins the hose.

When all is ready turn on the steam; go slow at first, till the hole is well entered; in this way it is started perfectly round and true and prevents cornering. Do the same when passing through a slanting seam, otherwise the steel will try to follow the seam, when it will bind and stick. If the holes should work out of line so that the steel sticks, tap the STEEL lightly with the wrench, but NEVER HAMMER THE PISTON OR CHUCK; if tapping the steel does not start it, shift the position of the machine slightly by loosening the back bolt, or otherwise, till it works free, then tighten the bolt.

Drills are all thoroughly tested before shipment and the parts are fitted closely; for this reason the drill may work a little stiff for a few days, and when run by steam the piston may stick at first (owing to the unequal expansion of the large and small parts and the condensed steam in the cylinder), until the drill becomes thoroughly warmed up and limbered. Blow the condensed steam out of the pipe and cylinder before giving it full pressure. In cold weather do not leave water in the cylinder; unscrew the stuffing-box and let it out to avoid freezing. If the piston sticks, push it back and forth a few times with steam turned on, till the water is forced out through the exhaust and the parts get thoroughly warmed; then oil all the parts. If the valve should stick, TAP LIGHTLY on the head of the bolt running through the steam-chest. NEVER POUND ON THE STEAM-CHEST. Run the drill slowly until thoroughly acquainted with it. As the bit cuts the rock the drill is fed forward gradually by turning the feed screw *slowly* and *regularly*

The speed of feeding depends on the rapidity with which the bit cuts into the rock. If fed too slowly the piston will strike the front head; if too fast, the piston stroke will be shortened; in either case the drill is expending the power upon itself instead of on the rock, because it is not being fed properly. The drill must be fed just enough to give the piston nearly its full stroke (after the hole is fairly started). To do this properly is an important point in running a drill, but it is readily learned by an intelligent man.

In starting a hole with a drill, the stroke can be shortened to about two inches by simply feeding the drill up close. This gives a short, light blow, making the starting of a hole in an oblique face easy and quick.

When the AUTOMATIC FEED DRILL is used, the feed screw is held fast by a pin placed in the hole in the back of the rest; then the drill feeds itself. The rapidity of the feed is regulated by tightening or loosening the friction straps which clamp the feed ratchet.

Do not bring too much strain on the feed screw and shell by running the cylinder out too far. If possible, keep water in the hole. If the holes mud up, clean them out when steels are changed. Use Kirk's water injector, page 84.

If the COLUMN is used, a piece of heavy plank or timber must be placed against the rock under each end of the column. The two jack-screws of the tunnel column should rest on the same piece of timber and should be jacked evenly.

Take care of your machine and use it properly, and it will last for years with few repairs.

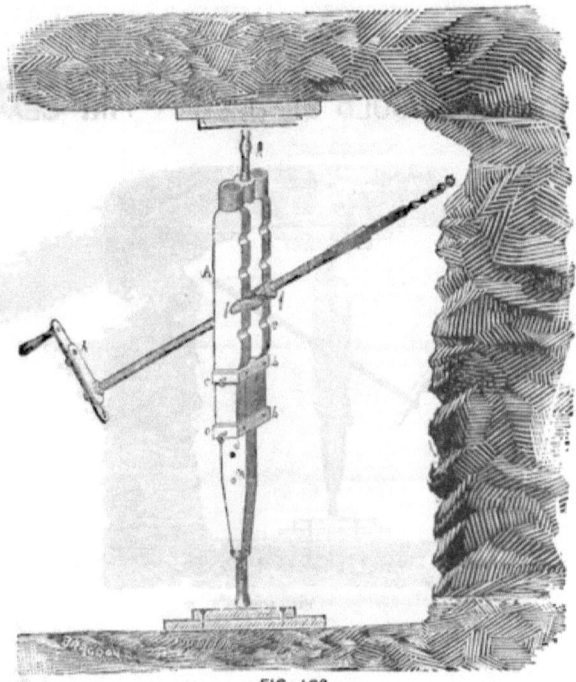

FIG. 103.

GRIM PATENT COAL AND FIRE CLAY BORING MACHINES.

Figs. 103 and 104 represent the large and small sizes of the Grim Patent Hand Coal and Fire Clay Boring Machines. They are very simple and easily set up. A is the frame, consisting of two pieces of oak, 4x1¼ inches, fitted into casting at upper end. This casting carries the tightening screw, K, and the oak pieces have notches, C C, up along one side to receive opening nut, F, and G is the boring stem with socket at G, to receive different lengths of augers, and the other end has the operating crank H. The lower ends of oak sides, A, are held together by wrought iron bands, D D, which carry the extension pins, C C. By withdrawing these pins, the extension timber, D M, can be shoved up in between the upper timbers, and the machine shortened to suit high or low places in the roof. When a hole has been bored the nut, F, can be easily opened, and the screw boring stem, G, can be just lifted back to starting end without unscrewing it the whole length. These machines are made in six sizes and bore holes from two to ten feet deep and from one and one-half to two and one-half inches in diameter. In ordering it is necessary to state the diameter and depth of the hole desired and length of post or thickness of seam.

POWER DRILLS SHOULD BE USED IN FIRE CLAY MINES.

FIG. 104.

Short Size Grim Drill for Low Coal or Clay Seams.

The above Hand Drills are very useful, and a great improvement over one man holding a steel and another striking with a hammer; but they are far behind what might be used to great advantage in fire clay mines now, just as the grain cradle was a great improvement over the sickle; but the grain cradle has no standing alongside of the grain cutter and binder, because all hand machines are limited to the strength and endurance of human muscles, in which the power is generated by high-priced fuel, consisting of fine roller-process family flour, beef, mutton, ham and eggs, pies, etc., while in machines the moving power is generated from cheaper fuel, and at less than on hundredth part the cost of above, and is applied by means of wood, iron or steel muscles, that never tire or relax their power. All we have to do is to find a suitable combination of mechanical powers to apply that power in a suitable manner to drill fire clay or perform any other laborious work we wish done, and when this is accomplished laborious work of any kind can be done far better and cheaper than by human muscle.

I have known of but two attempts having been made to use power drills in fire clay mines, and in both cases it was a complete success, but workmen combined against them and done everything they could to prevent their use, and for want of sufficient nerve in the managers of the works both plants were thrown out at great loss to owners.

From all I have ever seen I am sure that if the directions given on page 85 are followed, and then a range of continuous holes fired by electricity, fire clay can be mined for at least 50 per cent. less than it costs by hand labor.

COAL MINERS' DRILLS, SCRAPERS AND NEEDLES.

These very simple but generally used tools are now made so perfect and cheap by persons who make a specialty of them, and have the proper tools and machines arranged for making large quantities, that no jobbing blacksmith can either make them as good or as cheap as the persons thus prepared can.

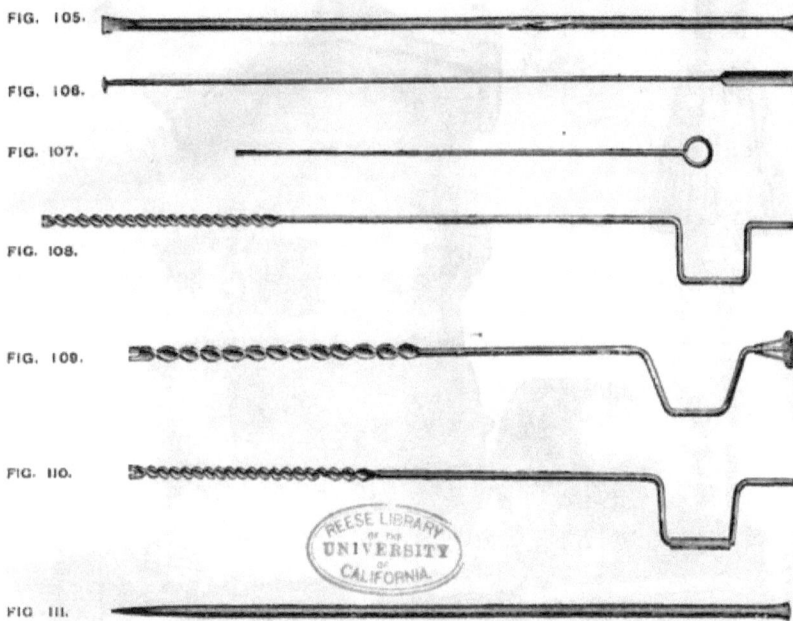

FIG. 105.

FIG. 106.

FIG. 107.

FIG. 108.

FIG. 109.

FIG. 110.

FIG 111.

Figure 105 shows a combination hand drill and tamper five feet long, one inch in diameter and can be dressed to make holes any diameter.

Figure 106 is a combined powder spoon and scraper four feet eight inches long, shank one-half inch in diameter, spoon one and one-half by seven and one-half inches, scraper head one and one-quarter inches.

Figure 107 is a miner's needle four feet long.

Figure 108 is a coal auger stem three-quarters of an inch in diameter, screw barrel two feet long, dressed to make holes one and one-quarter inches in diameter and five feet deep.

Figure 109 is a coal auger screw barrel two feet eight inches long, to make holes two inches in diameter and five feet deep, with breast plate attached.

Figure 110 is a coal auger screw barrel two feet two inches long, dressed to make holes one and one-half inches in diameter and six feet deep.

Figure 111 is a combined probe and tamp bar, one and three-eights inches in diameter. It is very useful in stump blasting; pointed end for probing among roots for place to bore holes with stump auger, as described in stump blasting, and the other end made a half inch larger, to use as a tamping bar.

ROCK AND ORE BREAKER.

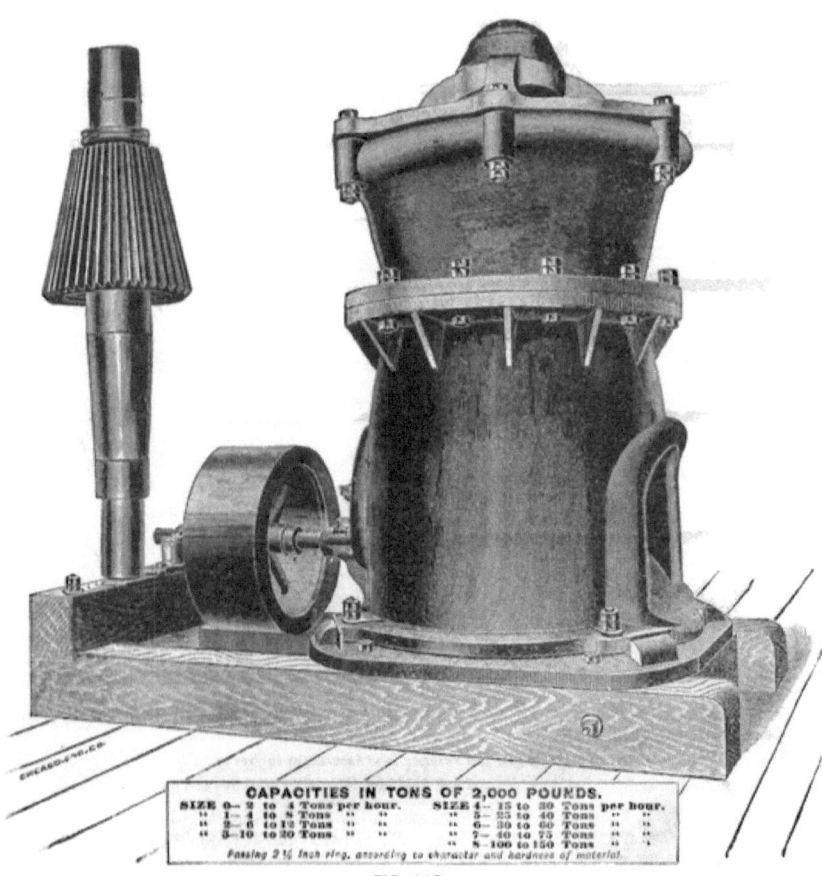

FIG. 113.

The above cut shows the best stone breaker in the world, mounted on strong, permanent timbers, ready to be made fast to foundation. The cut also shows the strong shaft, G, standing outside the shell with pulverizing cone, F, attached; explained on page 111.

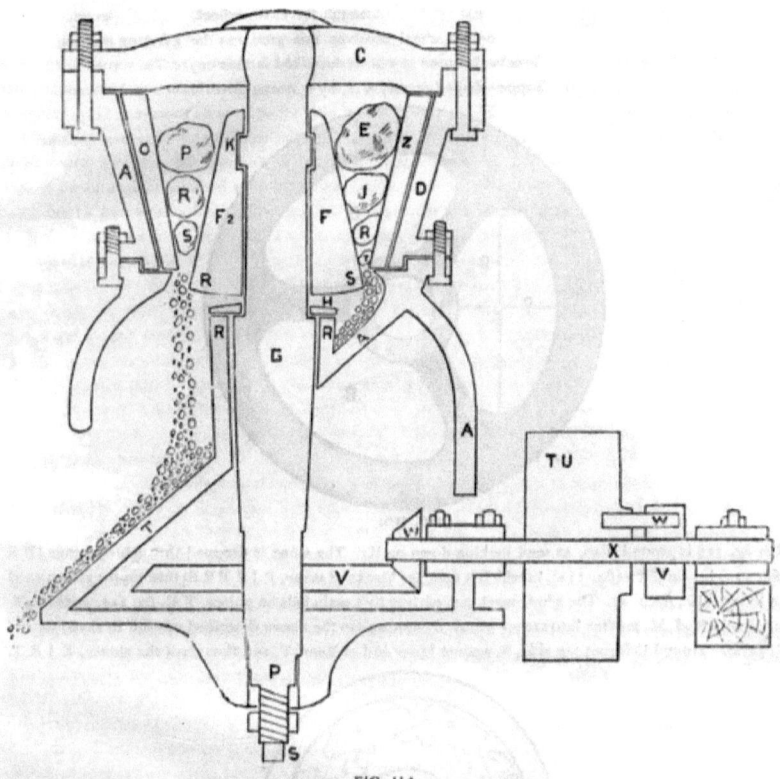

FIG. 114.

The above cut, figure 114, shows a sectional view of the best stone breaker I know anything about for putting out a large quantity of broken stone of any desired size, with very little power in proportion to the work done. It can be changed for coarse or fine work very easily by simply turning screw S to the right or left hand, which will close or enlarge the openings R and S.

THE GATES ROCK AND ORE BREAKER

Has a continual revolving motion, is very strongly made and has very few parts to get out of order or keep in repair. It consists of a very strong cast iron frame (A A A, fig. 114), and a strong forged steel shaft up its center (see G, fig. 113), which carries breaking head (F, fig. 114). This shaft, G, is supported on chilled cast iron step (P, fig. 118), and passes loosely through the crown wheel, L, a little to one side of true centre of crown wheel, L, and as crown wheel revolves this produces the gyrating motion of the shaft, G, and break head, F (which breaks the stone as will be described farther on). The top end of shaft, G. is held in position in center of hopper-shaped frame, A A, by a strong three-branch spider frame, C, of

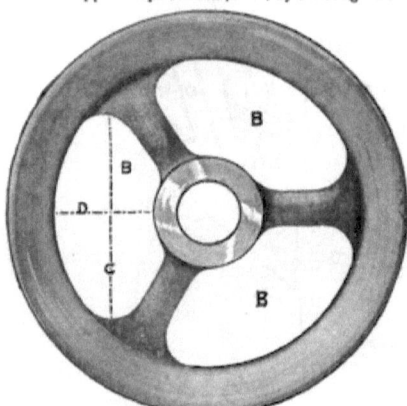

FIG. 115.

which fig. 115 is ground plan, as seen looking down on it. The stone is dropped through openings (B B B, fig. 115) into hopper (fig. 114), in which is seen the blocks of stone, E J R P R S; this hopper is furnished with liners (O Z, fig. 114). The whole machine is driven by a main belt on pulley, T U, fig. 114, on shaft, X. Spur bevel wheel, M, meshes into crown wheel, L, which gives the above described motion to shaft, G, and as it gyrates around it forces the side, S, against lower end of liner, Y, and thus gives the stones, E J R T,

FIG. 116.

a squeeze between the break head, F, and hopper, D, and while the full side, G, is forced against Z D, it at the same time draws, F2 R, away from side of hopper, A, and permits the stones, P R S, to drop down in the wedge-shaped chamber between A and F2, and so on continually all round the hopper. Another peculiarity of this machine is the combination of the convex hopper (A D, fig. 115), with the corrugated cone, F F2, as shown in fig. 116, where the stones, 1 2 3 4 5 6, are all pressing in the center against circular break head, F, and their extreme ends on opposite sides are pressing against the hopper, A D, thus placing every stone in the most favorable position for breaking, and all broken fine enough to go through the openings opposite points R S, drops at once on the inclined shoot, A T, and thus passes out of the machine, leaving only the larger pieces to be crushed again and when broken the fragments require no handling, but of their own weight drop down, while the break head, F F2, fig. 114, is pressed against the opposite side of hopper, A D, to a narrower place of hopper, A D, and still pressed against breakhead, F F2, ready to receive another squeeze when shaft, G, comes round, and so in a few revolutions of shaft, G, any sized stone that can be gotten into the machine is small enough to drop through the openings, S R, on to shoot, A T, and from there on to a pile or into a revolving screen or bin as desired (see fig. 128). Another peculiarity of this crusher is its peculiar construction. It provides for protecting its bearings and wheels from all dust or grit from the crushing of stone. Fig. 114 is a cross section, as if the machine was cut in two vertically, and snows the shaft, G, passing through a strong boss in the shoot, A T, with a tight fitting collar, H, fitting tight around shaft, G, but sliding easily on top of boss, R R, and preventing all dust or grit getting in on pinion, M, meshing into gear wheel, L, or into the bearings of shafts. The output of this machine is only limited by its size. The No. 8 machine will easily put through one hundred and twenty-five tons every hour it runs, if kept supplied with stone fast enough. To make the output coarser or finer, use the screw, S, fig. 114, below bottom of shaft, G, by screwing it so as to raise shaft, G. It will contract openings, R S, and so make the output smaller, and by screwing it down it will make the output coarser. Another great advantage in this machine is, that the band wheel, T U, fig 114, is a loose fit on shaft, X. The break pin hub, V, is keyed fast on shaft, X, and has a hole in it through which is passed the break pin, W, into a hole in the hub of band wheel, T U. The break pin is held in place by the set screws in the break pin hub, V, and is of no more than sufficient strength to withstand the strain necessary to break the stone in the crusher, but should a piece of steel (like dropping a sledge) get into the crusher, the strain would be so great upon the brake pin that it would break off and let the band wheel revolve free on shaft, X, while the machine would stop without breaking until the obstruction had been removed and a new break pin put in. The loose collars, H and I, are to keep the dust out of the journals and gyrate with the shaft, G, in which they fit tight.

OF RENEWING HOPPER LINERS.

The upper part (A D, fig. 114) is hopper-shaped and is lined by twelve chilled iron liners placed inside the shell, A D ; the space behind and between liners and shell is run full of melted zinc. When these liners are worn after long use and need to be changed they can easily be removed by driving in the key liner (which has reverse bevels on its edges) by the use of a wrought iron or steel pin through the hole in the shell, A D, at 2, fig. 118.

FIG. 117.

Fig. 117 represents a very strong overhead traveler, used for carrying the block and falls when setting up or repairing the larger size breakers. This is a most useful tool, and every breaker should have one, as it will be found a time and labor-saver.

DIRECTIONS FOR SETTING UP AND CARE OF BREAKER.

The breaker is furnished on hardwood frame, as shown in fig. 113. It should be bolted at each corner to foundation with one and one-half inch bolts for the larger sizes and one inch bolts for the smaller sizes. The foundation may be of masonry, substantial wood frame, or, if set down on the ground, simply cross timbers let into the ground and bolted to them. If the machine is shipped ready to run, follow directions for starting and oiling as given below. If upright shaft, G, is not in when breaker is shipped, before you put it in be sure to see that the eccentric, D, is perfectly cleaned out, and two to four quarts of the breaker oil put in, and see that the coil of hemp packing is perfectly fitted round the shaft inside the collar of bevel wheel, L, as this packing is intended to hold the oil and not let it pass through the journal too fast. Put the collars, H and I, into place before you put in the shaft. Put on the top, C, drawing it down evenly with the bolts, so it is perfectly tight, and using calipers or a blunt wooden wedge, so that the space between the lugs on the shell and those on the top shall be exactly the same on all sides. If the shell is divided into two parts, Q and QQ, and the upper half is shipped separately, draw them down evenly with the bolts until they come iron to iron, and proceed as above. If the frame with bottom plate, 3, is shipped separately, put on the lower half, QQ, and draw it down evenly with the bolts iron to iron, and drive in the wrought iron wedges between the lugs on the bottom plate, 3, and the shell, QQ, then proceed as above. If the shell is in one piece, and is shipped separately, it should be drawn down on bottom plate, iron to iron, and wedged as above. In making iron to iron joints it is a good plan to put in little strips of very thin paper at several points around the shell, and when none of them will pull out you can be sure your joint is right. Be sure always to have wood or leather liners in the boxes for counter shaft, X, as the bolts will not stay tight if there are no liners.

FIG. 118.

Sectional Perspective View of Breaker.

OILING.

The most important thing to be done is to keep all the bearings well oiled with best lard oil, and a little grease on the gear occasionally. We have made expressly a very heavy oil for the eccentric bearing, D, only, which is still better than lard oil for that bearing. The bearing at top of shaft, G, needs oiling from four to six times a day when running steady, only a little at a time. The boxes—for hand-wheel shaft—oil every two hours, or oftener if they get warm. The journal for eccentric, D, at lower end of shaft, G, is the most important part of the breaker to be looked after. Before you start up be sure to fill this journal or reservoir with oil, and after pouring it in, examine it a second time to make doubly sure, as it will sometimes run over before it is full, and may deceive you. It will take from one to six quarts, according to size of breaker. The oil may be poured in by raising the loose collar, I. After this first oiling, and when the breaker is first started, watch this eccentric journal very closely. You can put oil in through hole, J, in loose collar, I, and be sure to use plenty of oil. There is a pipe with plug in the bottom plate, 3, to draw off the oil from the eccentric when it gets bad from use, and by running hot water down this journal and out through pipe it will work out all the dirt and grit, then fill up with fresh oil, as when you first start. This oil should be drawn off every day or so the first week, after that, say once or twice a week. The oil you take out every day the first week will not be very dirty, and may be used probably on other parts of the breaker. Fill the oil cup on hand wheel, TU, once a day, so that the shaft will not be cut by wheel revolving upon it should the break pin be broken.

Whenever you wash out the eccentric, D, journal, first take out the hemp packing around the shaft, G, in collar of bevel wheel, L, so that your hot water will have a good chance to get into all the little holes, and wash everything out clean. Be very liberal with your water, and have it very hot. Wash out the hemp packing in very hot water, and coil it back inside the collar on bevel wheel, L.

STARTING.

When the breaker is first started, the shaft, G, should be lowered down as far as it will go, then set up the lighter screw, S, just so it will carry the full weight of shaft, but not enough to raise the shaft any. Now, if it does not crush fine enough with shaft in this position, then keep raising the shaft, G, by means of lighter screw, S, till the bottom of head, F, is even with bottom of concave, E, and no higher In this way you will break a half inch finer than when shaft is clear down. If you want to break still finer, then you will require thicker concaves, a larger head, or a narrow concave, so they must be set in, as shown in fig. 114. Endeavor to have such a size of head and concave as will give the right size opening at bottom of head, F, and concave, E, to produce the right size of material when the head is clear down. When raising or lowering the shaft, G, run the breaker light at first till it gets its new bearing, else the eccentric may get hot. After you have shaft in right position put some hemp packing with white or red lead around the screw, between bottom and jam nut, R, and screw up the jam nut perfectly tight to prevent oil leaking out of eccentric bearing. Put one man on at first to feed the breaker, and feed very slow, watching the eccentric journal to see that it does not get hot. If it heats, stop, let it cool off, put in more oil, and proceed in this way the first day or so, and gradually increase feeding as you can without heating, till bearing gets worn smooth, as long service will depend somewhat on starting right. After you have been running a day or two you should take up all the slack in the nuts on the shell, and also the wedges, and continue this at intervals of a day or two until there is no more slack. This same precaution should be taken with the top, C, every time you remove it. These joints can never be put together the first time so firm that the running of the breaker will not loosen them, hence it is necessary to see that they are taken up until perfectly solid.

Never run the breaker to exceed two or two and one-half hours without giving the eccentric, D, journal fresh oil, putting same in through the hole, J, in loose collar, I. This is imperative.

PLANS FOR CHANGING PARTS OF BREAKER.

FIG. 119.
Plan for Setting Concaves.

1 Is place to run zinc behind concaves.
6 Is concaves.
15 Is key concave, which must always be set opposite hole in shell. (See fig. 114).
4 Is wood or iron ring.
5 Is wedges for holding concaves tight to shell.

FIG. 120.
Plan for Setting in Concaves so as to Break Finer.

1 Is place to run zinc behind concaves.
2 Is hole in shell for driving out key concave, and the key concave (15, see fig. 31) must be set in opposite this hole.
3 Is pieces of iron set between concaves and shell, and opposite the lugs on back of concaves.
4 Is wood or iron ring.
5 Is wedges for holding concaves tight to the piece of iron and shell.
6 Is concaves.

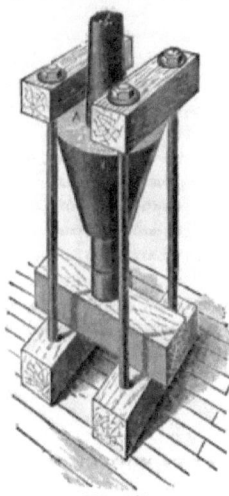

FIG. 121.
Plan for Taking off Head.

A Is point to use sledge when full pressure is on screws, also use sledge at opposite point on head at same time.

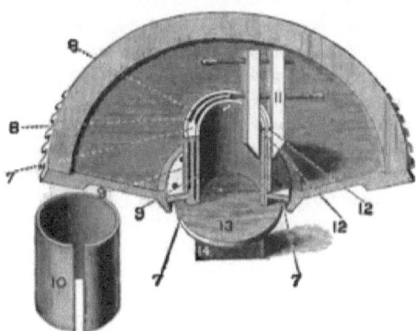

FIG. 122.
Plan for Babbitting the Eccentric.

7 Is the brass eccentric.
8 Is places to run in the babbitt.
9 Is inside and outside sleeves of cast iron.
10 Is one of sleeves 9 before fitting to eccentric.
11 Is screw clamps for holding sleeves tight to thin part of eccentric.
12 Is opening or slot in sleeves.
13 Is board to prevent babbitt running out.
14 Is block for holding board tight to eccentric.

REVOLVING SCREEN.

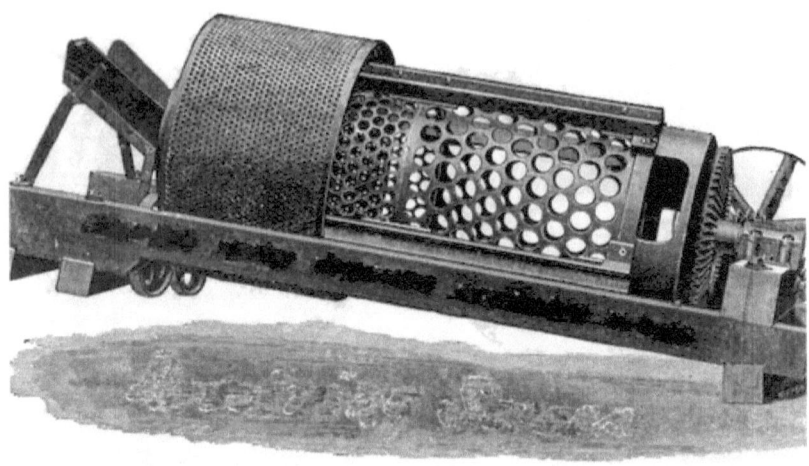

FIG. 123.

Fig. 123 represents an improved revolving screen and dust jacket attachment for separating different sizes of broken stone into bins—a convenient arrangement for making roads where they require coarse broken stone for bottom and finer on top. The screen is adjustable; one end can be raised or lowered to discharge slow or fast, if required. By taking off one rib any section of screens can easily be taken out and a different size put in. These screens are made in various sizes, and are the most perfect machine for the purpose made. Hundreds of them are at work, all giving perfect satisfaction. They are an indispensable adjunct where graded sizes of stone are required. The object of the dust jacket is to increase the capacity of the screen, that is, to save expense and make a short screen as far as practicable do the work of a longer one. It is especially valuable where more than three sizes of stone are required in a short screen. In this arrangement the first size larger than the dust section becomes the first section of the main screen ; thus what passes through that section and the dust only drops on to the outside dust jacket, and the wear and tear of all the other sizes which pass over the dust section in the ordinary arrangement is saved in this case. They may be run from a counter shaft, or by a back gear and belt to run them direct from breaker. They are made both with and without the outside dust jacket.

ELEVATORS.

For the economical handling of broken stone there is nothing more practical than elevators. These elevators are made in all lengths and of capacities suitable to the various breakers. They are all built from special designs ; are simple, durable and economical. They are largely used to elevate the broken stone from the breaker to bins. They may be run from a counter shaft, or gear and belt to run them direct from breaker.

PORTABLE BREAKER.

Fig. 124 represents a breaker in portable form, for street or other work, where it is necessary to move often, with elevator, and also a screen for separating dust and dirt from broken stone when required. The elevator, or screen, or both, can easily be detached. When you do not want to separate the dust and dirt from the broken stone the screen will not be required.

Only sizes 1, 2 and 3 are thus mounted. The framework is made of best seasoned white oak and is very substantial, and no expense is spared to make this plant capable of withstanding the heavy strain it is subjected to.

FIG. 124.

119

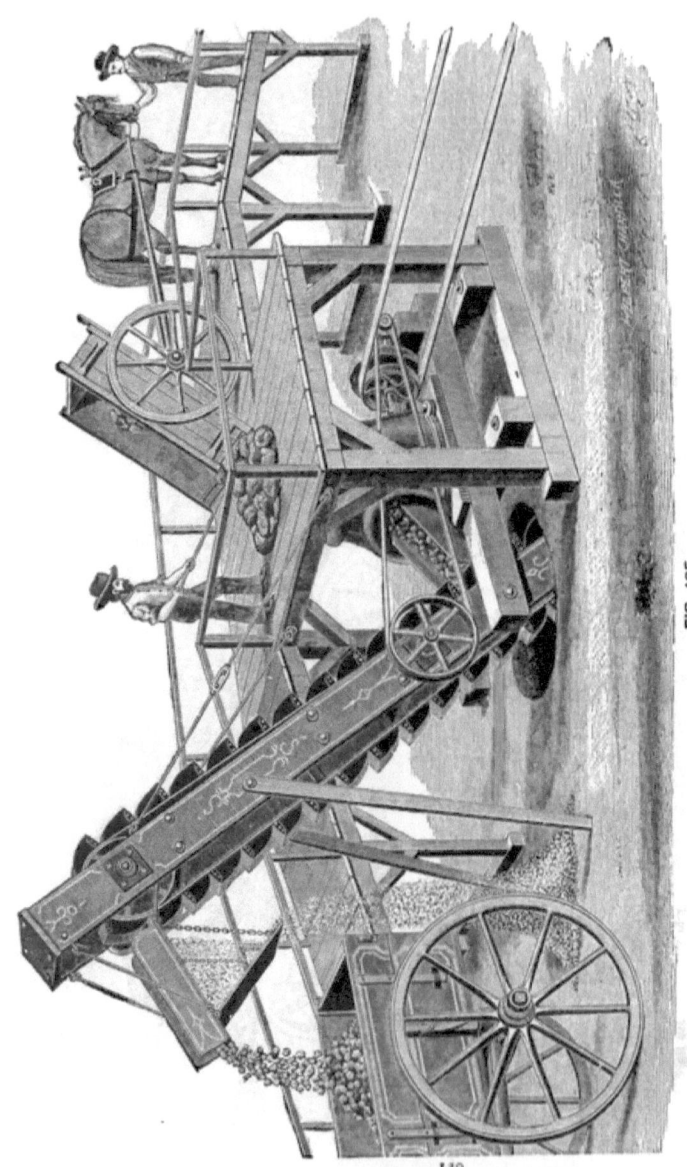

FIG. 125.

Fig. 125 is plan for setting breaker on any quarry where surroundings are all level. When breaking large amounts of stone the better way would be to erect a bin at end of elevator and discharge into that, then spout from the bin direct into wagons. In this way you could carry a supply of broken stone, so as to keep breaker going steadily in case teams were not there on regular time, or at all times; and if desirable to separate the dust and dirt from broken stone, make a division in the bin for that purpose; or, if supplied with plenty of teams and help, do away with the elevator and screen and discharge upon the ground, and have men there to load the wagons. It will be a great saving, however, to use the elevator and bin.

FIG. 126.

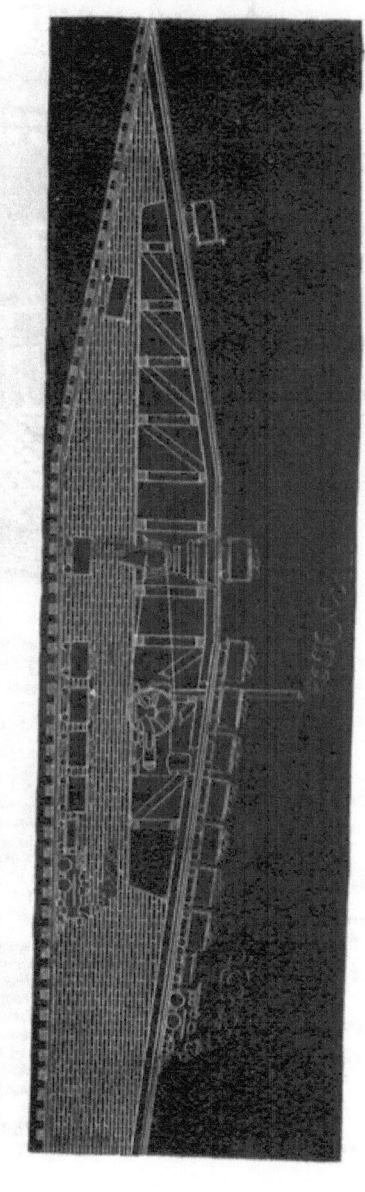

Fig. 126 represents one of size 6 breakers, as set and operated by the Baltimore & Ohio Railroad, at Elk Ridge Landing, Maryland.

It is one of the best plans ever adopted for handling large quantities of rock with very little manual labor.

While they have never made an effort to test this breaker, it has made a respectable record, as the following extract from a letter by W. G. Primrose, Esq., master mechanic of the Baltimore & Ohio Railroad, will show:

BALTIMORE, MD., July 13, 1882.

GENTLEMEN :—On the 18th of May I put the babbitt box in our size 6 stone breaker. It has been breaking on an average six tons in seven minutes, and not forcing it.

Yours truly,

W. G. PRIMROSE.

LATER.

121

FIG. 127.

Fig. 127 represents the plant of the Cambria Iron Works, erected at Birmingham, Pennsylvania. This breaker is the size 8, the largest built. The letter of Mr. John Fulton, the general manager, will convince any one of the merits of this mammoth machine.

IRON AND STEEL WORKS OF THE CAMBRIA IRON COMPANY.

E. Y. TOWNSEND, President
POWELL STACKHOUSE, Vice President.
JOHN FULTON, General Manager.
                                        JOHNSTOWN, PA., Nov. 2, 1888.

DEAR SIRS:—Your favor of October 31st received. I visited the Birmingham limestone quarries of the Cambria Iron Company yesterday, and found the large breaker (the mastodon of its class) working with the utmost ease. The machine is breaking five hundred tons per day, which appears to be less than one-half the work of this powerful breaker. We are extending the quarries to get the product up to seven hundred or one thousand tons per day. The breaker does its work very easily, and we estimate that when we get up to seven hundred tons per day it will cost us about two cents per ton to break the rock. The limestone at this point is rather tough, and requires more force to break it than the limestone near your city.

I may say we are very much gratified with the results of this large breaker. You will remember that, when Mr. Stackhouse and the writer accompanied you out to the limestone quarry south of your city, we saw, I think, two or three men keeping the breaker open, and it was for this reason that a larger breaker was suggested. Its economical working shows the wisdom of this increase in its capacity, and I am not sure that you will not be urged to make a larger one still.

A number of persons have been down to see this breaker at work, and they all return entirely satisfied as regards its ability to do what is claimed for it. Without urging it very fast, it will break two tons per minute readily. I think any reference that may come will receive full satisfaction in every respect in regard to the capacity and economy of the machine.   Very respectfully,

                                        JNO. FULTON, *General Manager*.

May 28, 1891.—And now after using the above stone breaker two and a half years every word of the above has been more than justified.

122

FIG. 128.

Fig. 128 shows breaker, elevator, screen and bins, in which to store broken stone when not needed immediately. These bins are of great advantage along with a crusher, as they permit the quarryman to work on, and then when broken stone is wanted in a hurry it can be delivered as fast as needed. This plan is adapted for a flat, level country.

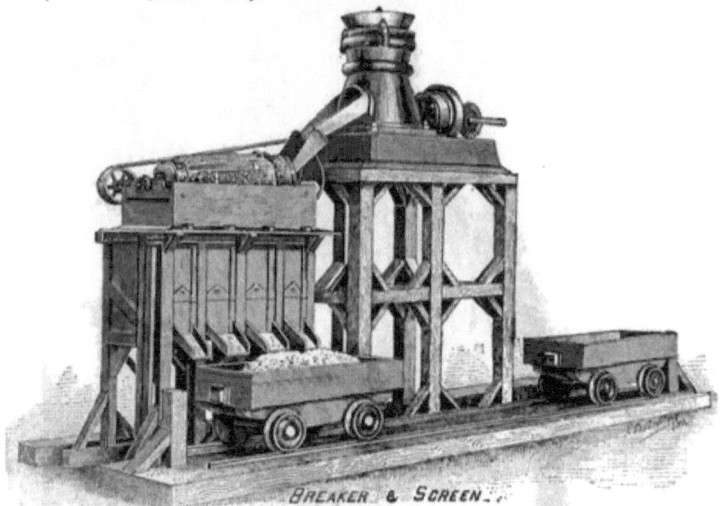

FIG. 129.

Fig. 129 shows a plan adapted to a hill-side, where the stone can be put into the breaker from above and from that discharged into the screen without an elevator.

FIG. 130.

Fig. 130 shows a plan for a location, where the stone is located up in a hill, but railroad for shipping broken stone on is so located that it would require a very high trestle work to get the broken stone into breakers and then into bins. It also shows a double-sided bin, where cars can be loaded on either side.

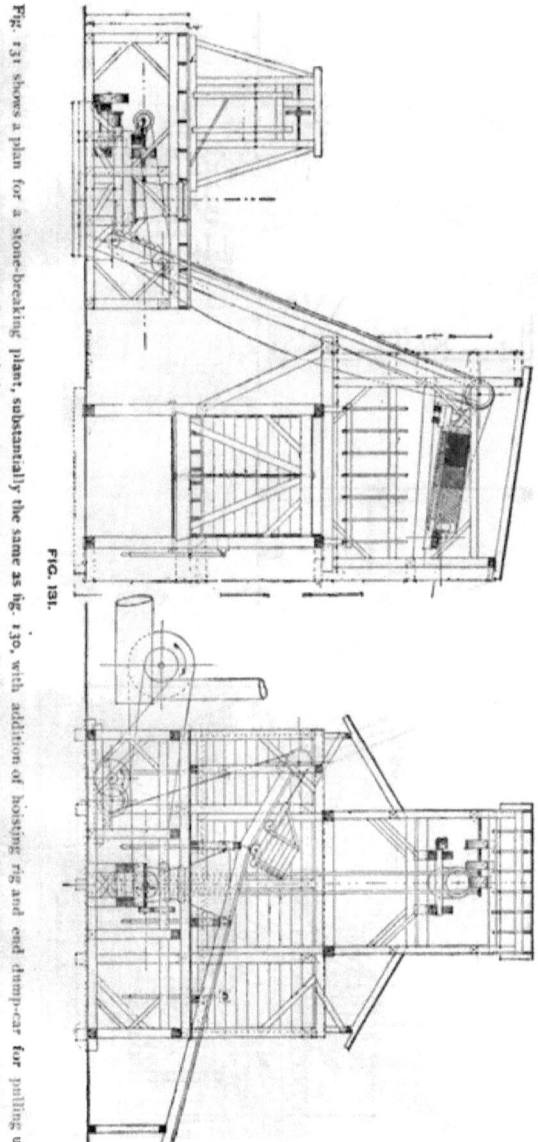

FIG. 131.

Fig. 131 shows a plan for a stone-breaking plant, substantially the same as fig. 130, with addition of hoisting rig and end dump-car for pulling up stone from the quarry to the breaker, and elevator discharging into a screen. This plan also shows form of house for covering machinery.

125

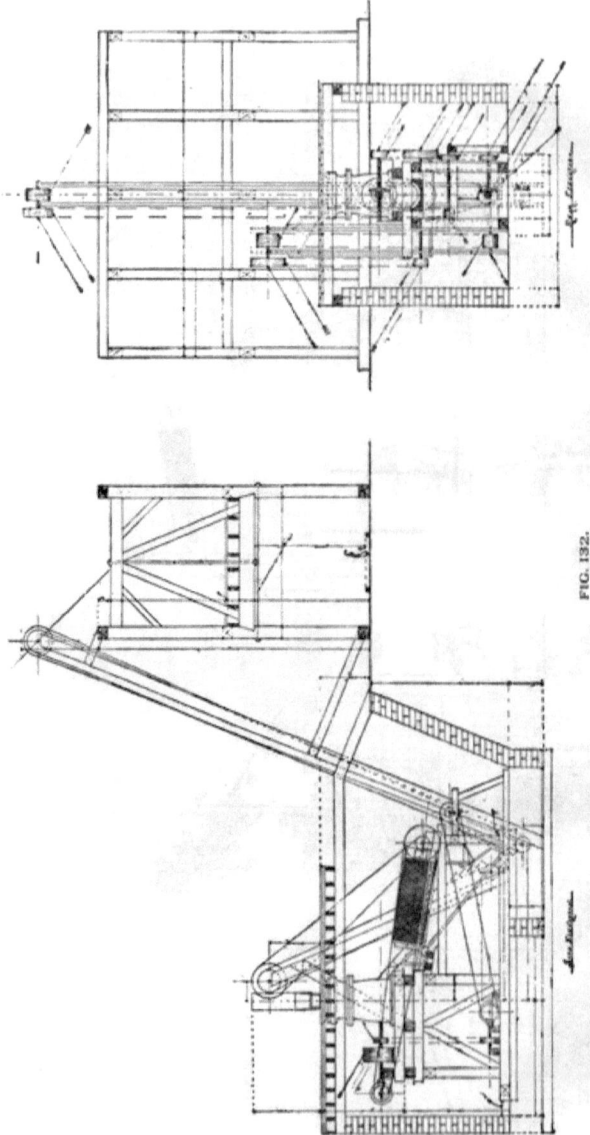

FIG. 132.

Fig. 132 represents a stone-crushing plant where large capacity and fine product are required. In this plan the breaker and the bulk of the machinery are set in a pit in order to bring the top of the breaker to the ground level. This plan is largely used in crushing granite for granitoid where the material must be crushed so as to all pass a ¾ inch hole or perforation. The operation is as follows:

The stone when dumped into the breaker is crushed and falls down into the revolving screen below, this screen having small perforations, or holes, say 1, ¾ or ½ inch as desired. That portion which is sufficiently fine to pass through the screen falls into an elevator which elevates it into the storage bins, where it is ready for use. That portion of the material which will not pass through the screen passes down into a short elevator which elevates it up and into the crusher, where it is again crushed along with the new material which is being continually fed into the machine. With this arrangement parties can use the larger size of breakers for very fine work, giving them large receiving openings, great capacity and fine product.

126

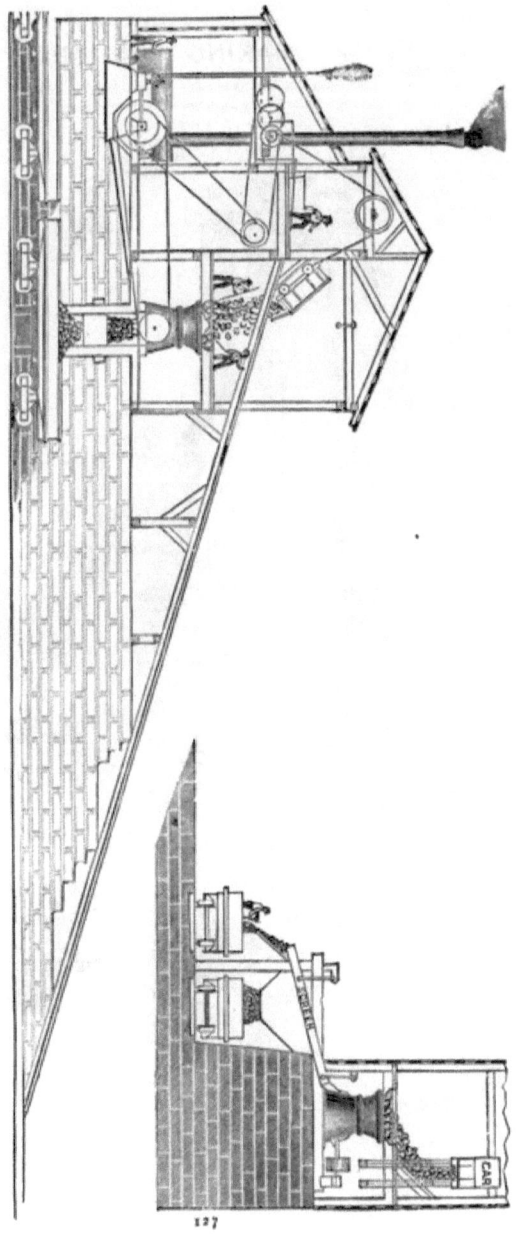

FIG. 133.

Fig. 133 represents a plant located higher than its supply of stone, where the same engine that runs the breaker draws stone from the quarries, thus saving all expense and trouble with horse or mule power.

127

PLANT FOR MAKING FINE MATERIAL.

It is also perfectly adapted to preparing certain ores for roasting, and other processes requiring an even product; for finishing material for concrete, top dressing, roofing, etc. It is claimed that this combination of breaker, to break the stone from the quarry (not shown in this cut), revolving screen to separate it to any desired size (see fig. 134) and the elevator to return any pieces too large to the breaker again, has never been equaled for quantity and quality of work done.

FIG. 134.

Fig. 132 represents a breaker with revolving screen and return elevator, and is designed to break the material very fine and to size it so it will all pass a certain mesh screen. With this plant it is practical to break so that everything will pass a ¾ screen. The great advantage of having product thus prepared for stamp mills, rolls or other reducing machinery is at once apparent.

ROAD ROLLER.

FIG. 135.
Latest Improved Pattern.

As will be seen in fig. 135, it is made in four separate rollers, thus securing great ease in turning, and these rings can be made any desired weight. But in case heavy rolling is wanted, as will be seen in the cut, there is a box before the roller and another behind into which a ton of stone may be put to add weight.

IMPROVED HOISTING AND CONVEYING APPARATUS.

Especially adapted for Quarries, Bridge and Dam Building, Contractors' Uses, Coal Handling, etc.—Description.

The engraving, fig. 138, represents an improved system of hoisting and conveying by means of a suspended wire cable. The cable may be either level or inclined, as desired, and is usually supported by A frames, and the ends securely anchored. The carriage travels either way on the cable, being propelled by means of an endless rope operated by a special type of engine, illustrated on pages 130 and 131. By means of this rope the carriage can be stopped and held at any point on the cable, while the stone or other material is being hoisted or lowered.

The hoisting rope is supported by means of the traveling carriers as shown. These carriers are indispensable where the span is of any length, and are fully covered by the Locke patent. It will be seen that in operation the hoisting and conveying may be carried on either separately or together; in the latter case effecting a great saving in time and in the former enabling a great degree of accuracy to be obtained in handling stone, thus making the apparatus of great value in erecting dams, bridges and other engineering work.

It is also in many cases well adapted for contractors' uses, especially where rock excavations are to be made. It is out of the reach of a blast, and will carry away the rock in large or small pieces, either by tongs or in scales, as shown on page 133, much faster than by any other method; and also saves breaking up the large pieces of rock. It will readily reach and drag a load at an angle of 45 degrees from the cable, and by using a snatch block this distance may be increased as desired.

SPECIAL ENGINE FOR HOISTING AND CONVEYING APPARATUS.

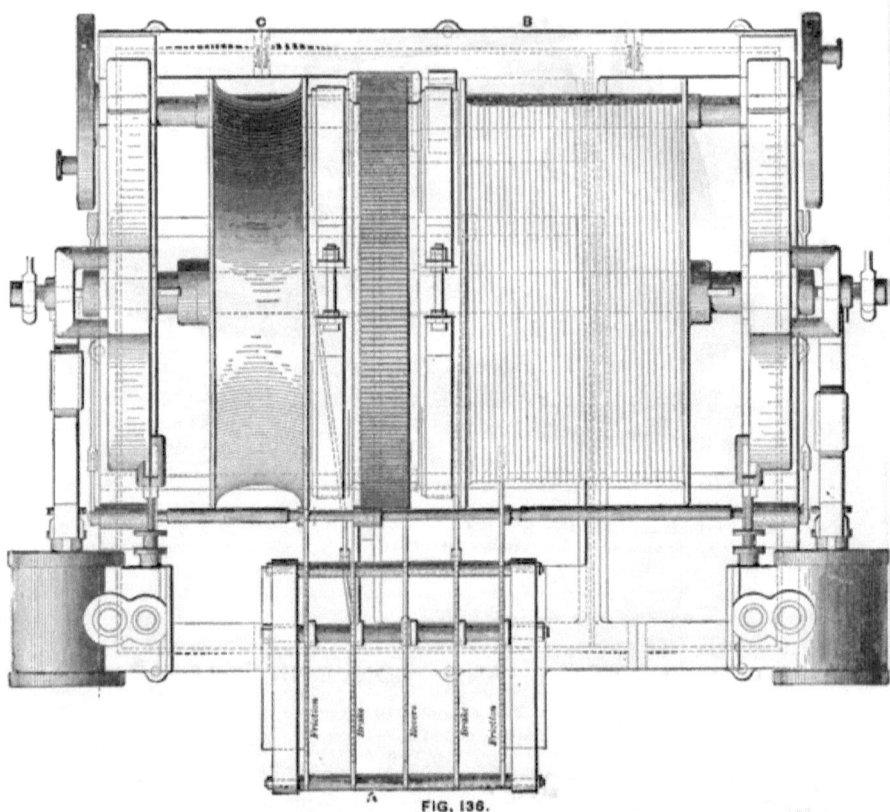

FIG. 136.
Illustrated on Opposite Page.

Figs. 136 and 137 show respectively a plan (or top view) and side elevation of an engine especially designed to operate the hoisting and conveying apparatus described on the peceding page. This engine has double cylinders with cranks connected at an angle of 90 degrees and is fitted with reversible link motion. The drums are of regular friction type, one drum, B, being spirally grooved to carry the hoisting rope, and the other drum, C, is turned off smooth with a curved surface, as shown in the engraving, and carries the endless rope. The endless rope is wrapped around the drum four or more times—enough to secure sufficient friction to keep it from slipping in the opposite direction to that in which the drum is turning—and the ends are passed over sheave wheels on the derrick frames (fig. 138) and made fast to the front and rear of the traveling carriage. This is the most perfect and convenient hoisting engine I have ever seen working; by means of it one man standing at A can handle more stone or dirt than twenty-five horses and carts can in the same time, and at less than one-tenth of the cost.

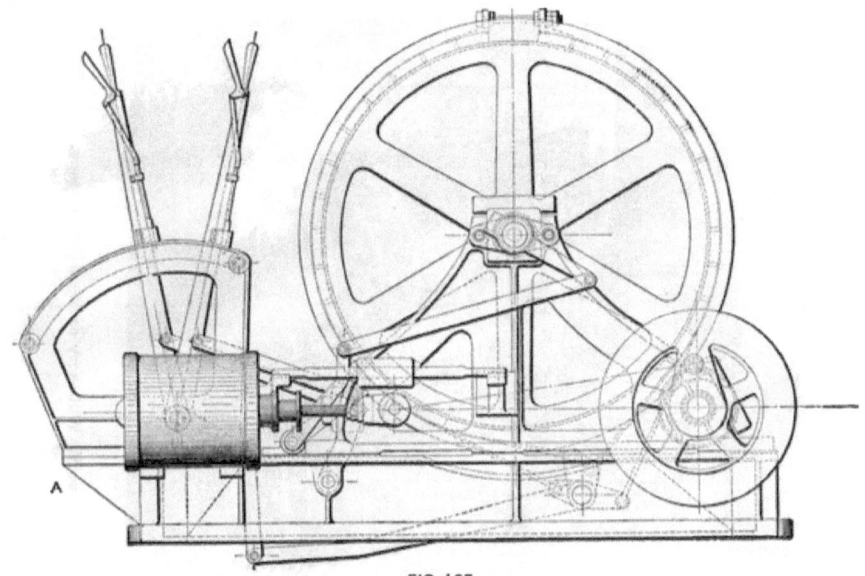

FIG. 137.
Side Elevation.

In operation the carriage is moved in either direction—the engine being reversible—and as the end-less rope on the curved drum, C, begins to wind the natural tendency of the rope to climb the drum is resisted by the curved surface which produces a slight lateral slipping of all the coils of rope on the drum. Thus the coils always remain nearly in the center of the drum, the slipping of the coils being gradual and almost imperceptible. The wear on the rope produced by this slipping is not appreciable; in fact, in a case where the ropes have been in constant use about a year the endless rope shows no more wear than the hoisting rope, both being in first-class condition.

This drum is provided with a powerful band-friction brake, applied by means of a hand lever, as seen in the above engraving, and is self-acting, locking the drum securely, so that the carriage is held at any point desired while the load is being hoisted or lowered.

The hoisting drum is perfectly independent of the other, and being of the same diameter winds at the same rate of speed and keeps the load at the same height if so desired. This drum also has a self-acting band brake, by means of which the load can be held positively. It will thus be seen that this inde-pendent action of the drums gives the operator perfect command over the apparatus, as he can use them together, or can hold either of them and use the other.

The reversing lever, friction levers and brake levers are all brought to a central position at A (fig. 136), so that the operator can work all of them without difficulty. The reverse lever quadrant has the usual square notches, while the other quadrants are serrated, and the levers having catches operated by thumb latches can be handled quickly, and will stay in any position in which they are placed. This saves a great deal of time in handling the engine, and adds to the general effectiveness of the apparatus.

These engines are made in three sizes of 30, 50 and 75 horse-power respectively.

FIG. 139.

Improved Single Wire Rope Hoisting and Conveying Apparatus. Will carry safely up to six tons.

FIG. 139.

Improved Double Wire Rope Hoisting and Conveying Apparatus. Will carry safely up to twenty tons.

COWIE'S PATENT

Absolute Prevention of Accidents from Overhoist with the Attendant Loss of Life and Damage.

Figure 140 shows a sectional view of a mine cage and slides—No. 12 the slides and No. 10 the cage, with its hoisting chains attached to upper end of cage and the top of cage, 10, which is very close to the top timber, into which the slides are mortised to keep them from spreading, and if by any neglect of the engineer or the derangement of throttle valve of the engine the cage should be drawn up too high, so as to make an overhoist or, in other words, to draw the cage so high as to force the top of cage so hard against the cross timber of the slides as to draw the end of the wire rope out of its socket (not shown in cut) on end of chain, or if the chain should break from being drawn up too high, then Mr. Cowie's improvement comes into use, but is not in the way or needing the least care until it is wanted, and when needed it is needed so bad that nothing else known to the human mind can take its place; for, if by any cause the cage is drawn up so high as to pull the hoisting rope free from the cage, then the cage with loaded wagon and load of coal, and the human beings which

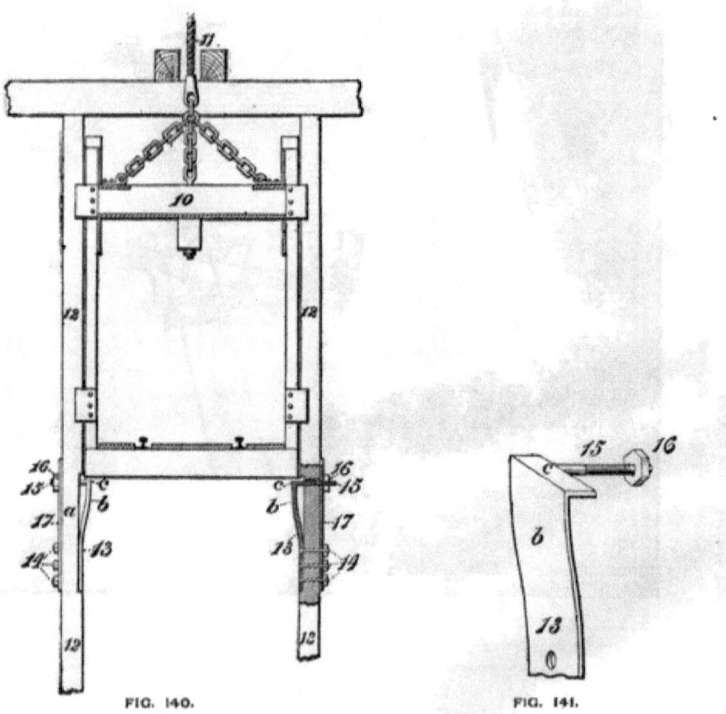

FIG. 140.   FIG. 141.

may be on it at the time (if it is not provided with this simple device) descends with all its load at nearly the velocity of a cannon ball to the bottom of pit; and the deeper the pit, the greater the velocity with which it will strike the bottom, completely destroying itself and every human being that may be on it at the time.

On January 4, 1889, the winding rope in the No. 2 shaft of the Connellsville Coke & Iron Company at West Leisenring, Pa., was broken by over-winding. The cage, containing three men, fell to the bottom of the shaft 500 feet below, crushing them into almost unrecognizable masses. The cage was smashed to splinters and extensive damage done to the shaft. Mr. James Cowie, Sr., at that time superintendent of shaft No. 3 of the Connellsville Coke & Iron Company, at Leisenring, Pa., began to consider means of securing a device that would effectually prevent all accidents involving danger to life and property by such over-winding, as had already resulted so fatally. After some weeks of labor, he perfected the device herewith illustrated and described. On July 9, 1889, he was granted letters patent, No. 406,630, on his device, and he now offers it to the consideration of owners and operators of coal and other shafts, and elevators of all kinds.

The inventor does not claim that his device will prevent over-winding, but that it will prevent accidents arising from over-winding. Several devices used to engage a falling cage, operating below the ordinary working course, have been made, but not until now has a device been perfected to operate above the working course so as to prevent a downward movement of a cage when a suspending rope may break. The simplicity of the device makes its action positive and at all times reliable. It cannot get out of order, and the shaft superintendent can always rest assured that he has an effectual check upon his engineer, no matter how careless that employe may be. Its cost is so trifling that no owner or operator of a shaft can afford to longer delay using it.

But all this can now be prevented by use of Mr. Cowie's device, which is so simple and plainly shown in the cuts as scarcely to need any description.

DESCRIPTION.

The device is simple in its construction and operation, as is shown in fig. 140. Figure 140 is the side view of a cage and its guides, representing the same as they appear when provided with the safety attachment. Figure 141 is a perspective view of a portion of one of the sustaining springs. In the drawing, 10 represents a cage and 11 the winding rope. The cage 10 runs between the guides 12, and the inner faces of these guides are each recessed at A. In these recesses the safety attachment is mounted, being bolted to the guides by bolts 14.

The safety attachment is simply a spring, the upper end of which is bent to form a reverse curve. Above the curves (shown at 6) the springs are bent so as to extend at right angles, thus forming shoulders, C; threaded shanks, 15, extend outward from the horizontal sections of the springs, and these shanks are engaged by nuts, 16, which bear against wear-plates, 17, which are secured to the outer faces of the guides, the bolts passing through these wear-plates.

The location of the safety attachment is such that the bottom of the cage will not ordinarily pass above the spring shoulders; but should the winding rope be over-wound, the cage will be carried above the springs. As soon as the bottom of the cage passes the spring shoulders the springs are released, and, in case the winding rope breaks, any downward movement of the cage will be checked by the safety attachment, and consequently all accidents from over-winding will be averted.

In case the winding rope is broken, and after it has been repaired, the springs, 13, can be drawn so that they will rest within the recesses, A, by turning the nuts, 16, as will be readily understood.

When one sees how cheaply and at the same time how effectively a reliable protection can be adopted to protect the lives of thousands of miners, it looks like barbarous indifference about the lives of others to run a mine shaft one day without it, and I see no reason why it should not be applied to every passenger elevator used in all our high buildings, in which hundred of thousands of persons are every day suspended over a pit, a fall down which would be certain death. True, all of these have protections against rope breaking below a certain height, but, as all human contrivances are liable to go out of order, it is better to have 90 per cent. of surplus protection than to be 1 per cent. short of enough.

WHAT IS THOUGHT OF IT.

The safety attachment is in actual use at the Monarch works, Leisenring, No. 3, of the H. C. Frick Coke Company. Rigid tests of the efficiency of the device have been made, and it has worked under all conditions of service.

Relative to the test, Mr. Robert Gray, Sr., superintendent of the works, says: "The safety attachment does all that is claimed for it, viz: To catch and firmly hold the cage if the rope is broken by over-winding or pulled out of the socket at the sheave wheel. We have thoroughly tested the device, and find that by its use it is impossible for the cage to fall back."

Mr. Fred C. Keighley, mine inspector of the Fifth Bituminous District of Pennsylvania, has examined the device, and says of it: "This improvement will be a preventative of that class of accidents which result from over-winding of the cages at coal mine shafts (in fact to any kind of elevators), and I have no hesitation whatever in recommending its adoption by coal operators and others."

Mr. Robert Ramsay, superintendent of the Standard mines of the H. C. Frick Coke Company, says: "I have examined the safety catch and believe it adds additional safety in cases of overwinding when the rope breaks by the cage coming in contact with the overhead timbers. We are about to put it in our Standard shaft."

Mr. J. H. Paddock, chief engineer of the H. C. Frick Coke Company, says: "We have one of Mr. Cowie's safety attachments at our Leisenring No. 3 shaft, and I am satisfied that its use will prevent loss of life and damage to property. We expect to use them at all of this company's shafts where it is possible to put them in.

Mr. Morris Ramsay, general superintendent of the Southwest Coal and Coke Company, has this to say: "The safety attachment is a sure safeguard against all accidents arising from over-winding. I recommend it as being able to perform all that is claimed for it."

All information regarding the right to use the device, or anything pertaining to it, will be cheerfully given. Address all inquiries to

JAMES COWIE, SR.,
Superintendent of Royal Salt Shaft,
Kanopolis, Kansas.

ROCK CRUSHER AND PULVERIZER COMBINED.

This mill is composed of two cylindrical heads or cups, B B, figs. 142 and 143, arranged upon the opposite side of a case, into which they slightly project, facing each other, and are made to revolve in opposite directions. The rock being conveyed to the interior of the case through the hopper opening at the top, is retained and prevented from dropping below the revolving heads or cups by a cast iron screen ; and entering, as it must, the heads or cups in revolution, is immediately thrown out again from each cup, in opposite directions, with such tremendous force that the rock from one cup, in the collision with the rock thrown oppositely from the other cup is crushed and pulverized, and the grinding which otherwise would be upon the mill, is transferred to the material, which is at once reduced to powder. The method of reducing rock by this process differs entirely from any other in use.

This mill is of extremely simple construction, being composed of only four elementary parts—a case, two hollow heads or cups and a screen—and is easily run and kept in repair.

The attention of all who are interested in the crushing and grinding of ores and other hard materials is called to the absolute originality of this invention.

The statements of product made are facts, not put too strongly, as all who have used the mills will testify. That so small and simple a machine should accomplish so much work seems incredible, but is nevertheless true.

An excellent mill for reducing hard substances is shown in the following cut for crushing and grinding ores, phosphates, etc. The illustration gives a view of the machine as it appears in operation:

FIG. 142.

The material to be ground is conveyed through the hopper at the top to the case, A, filling the case and the revolving cylinders or heads, B B, which, being put in motion, hurl their contents against each other with such power that the rock is at once crushed to atoms. The mill does not grind the materials, but simply furnishes the power that compels the rocks to grind themselves; consequently the hardness of the rock does not affect the result, as it acts upon itself.

FIG. 143.

Figure 143 shows the cups or heads drawn back to give a view of the interior of the mill, showing the cast iron screen, C, through which the material, as fast as ground, passes and falls into the hopper marked D. When necessary to reduce the rock to a greater fineness than the screen outlets allow, the coarser part of what leaves the screen is reconveyed to the mill by an elevator for regrinding; that which is already sufficiently fine being first removed by the usual apparatus adopted in milling.

A suction blower causes the air to draw strongly into the mill, and prevents the escape of dust.

The cast iron screen, C, is composed of small sections, and the worn parts are cheaply and easily replaced. The wear upon this screen is very slight, as it is always protected from the action of the rocks thrown from the heads or cups by a cushion of interposing material formed by the rocks, which always fill the case and cover the screen.

FIG. 144.

Fig. 144 shows one of the revolving heads or cups taken apart. It is composed of two pieces, one of which, E, a simple hard iron cylinder (called the bushing) is removable, and when worn is easily taken out and replaced.

FIG. 145.

Fig. 145 shows the same put together and Z, the conical cup-like stone lining which, curiously enough, always forms itself inside the head or cup in the process of grinding. This lining forms itself by the caking within the cup of the material being ground, and completely protects the metal of the cup from wear, except at its edge. (See cut above).

It will be seen that, with the exception of the edges of the cup bushings, the entire interior of this machine is completely protected by the rock itself. These bushings are, in practice, very slowly worn and cheaply and easily replaced. They can be cast at the nearest foundry.

The following cut, showing the elementary parts of the mill in cross section, may possibly explain more clearly the operation of the machine when made to run :

FIG. 146.

Let B B represent the two opposite heads or cups of the mill, holding the two bushings, E E, which slightly project into the case. At Z Z the stone hollow cones are shown (which form themselves in each head by the packing of the rocks being ground, after the machine has been run a few moments). The hopper is shown filled with rocks, which drop into the case of the machine between the two heads ; let the mill now be started up. In a few moments the two stone hollow cones, Z Z, as before stated, form themselves and become as hard as the rock. When these hollow cones have formed, it is plain that the centrifugal force given by their revolution will hurl out of the hollow cones in the general directions indicated by the arrows all the rocks that are forced into them. These rocks thrown violently out of the two hollow cones, oppositely, cannot strike the case, for they are thrown from it and against each other, and the force lost by collision ; also, every movement of these projectile rocks is made through an atmosphere of the same material, for the case is kept constantly full of rock. Thus in the collision of the rocks the material is pulverized. If the flying rocks could take any direction not indicated by the arrows, before striking the case, such projectiles would have to meet and crush all the rocks lying in their way to the case, or screen, and they would be pulverized themselves or lose their force before striking the iron boundary.

This machine is excellent for work requiring the product to be reduced to dust, and as I have seen it repeatedly in operation I strongly recommend it for that purpose. **It requires the Stone Breaker shown on pages 110 and 111** to reduce the quarry stone small enough to enter the hopper on case A, fig. 142, and this, with the two machines combined, screen and elevator, shown in fig. 134, will enable any one to rush through a very large amount of fine, pulverized limestone for glass flux or other similar use at very little cost and very few repairs.

DITTMAN'S PATENT HAND-POWER ROCK DRILL.

Drilling capacity in Sand Stone, six feet per hour.

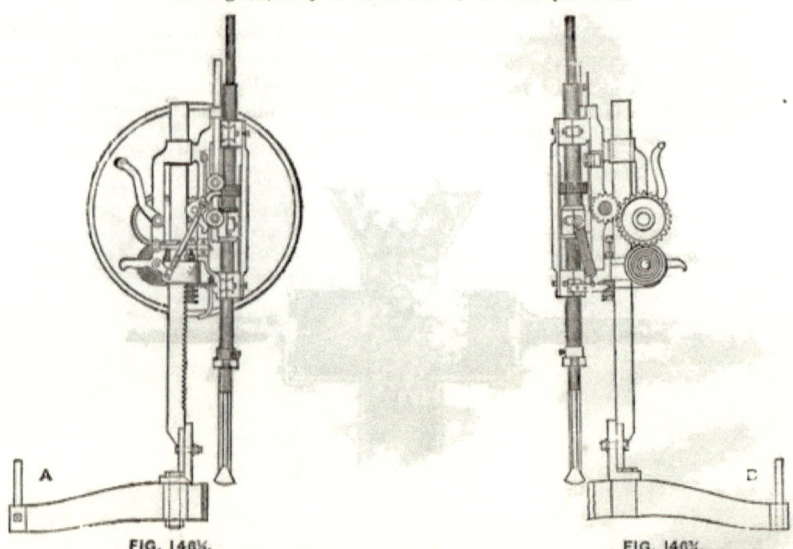

FIG. 146⅜.   FIG. 146⅜.

This hand-power rock drill has but recently been patented and put on the market, and appears destined to fill a long-felt want for something better than hand drilling in the old style, and is not so expensive as a steam or compressed-air plant. It has very few and simple parts to go out of order, and by it holes can be drilled to any desired depth for rock blasting. It is easily moved from hole to hole, and by means of set screws, A B, it can easily be adapted to any unevenness of the rock. It is self-feeding and self-rotating, and one man is all that is necessary to operate it as above.

PORTABLE AND STATIONARY POWDER MAGAZINES.

All Blasting or Sporting Powder should be laid down flat, whether in hauling or in storing.

For many years I have observed the great need of better arrangements for taking care of explosives about quarries and railroad work, and lately I have had the matter forced to my attention by seventeen kegs of best blasting powder being found spoiled for want of proper storage conveniences, and I therefore take the opportunity of calling attention to what I have said on this subject on pages 8 and 9; also, to add that I have observed that powder is almost the first thing needed about a railroad contract or opening a quarry, and it is generally the last article for which proper storage is provided. It is no uncommon thing for everything else needed on a railroad contract to be ordered and a beginning

FIG. 147.

made, and then comes a hurry for explosives. When the powder arrives there is no separate place to properly store it, but it is hurriedly put in a corner of the blacksmith shop or an old leaky-roofed stable, and soon a complaint is made that the powder is bad; but the user (or powder monkey) does not report that it has been treated as if it was limestone—not affected by the weather or surroundings. Now, to supply this long-felt want, the firm of Arthur Kirk & Son have made arrangements by which they will be able to supply portable rat and rain-proof sheet iron magazines on wheels, as shown in fig. 148, and capable of holding 16, 24 and 32 kegs; all these three different sizes will have rain-proof covers fastened to one side with strong strap hinges, and the other side with two strong padlock hasps and strong handles on each end, all painted outside and inside. The above firm will also supply stationary sheet iron magazines

FIG. 148.

like that shown in fig. 147 in four different sizes, capable of holding 100, 200, 300 and 400 kegs respectively; they are substantially made on a frame work of angle iron covered with strong sheet iron and made in sections held together at the corners by bolts inside, which can easily be removed and the sections can then be shipped separately thereby securing a low class of freight from one contract to another. See figs. 147 and 148.   Powder kegs should always be laid on their side and never on end, so that any water or moisture may run off.

THE FORSTER ROCK BREAKER.

Fig. 149 represents the Forster Rock Breaker and shows the construction of that machine very well. The power is applied by means of a belt on the tight and slack pulleys, shown in the center of the cut, which are mounted on a short driving shaft with a center crank and short pitman connected with the long lever reaching forward to front of machine and oscillating on a crown bolt under the ring shown in cut.

FIG. 149.

The end of this lever forms the end of feed box where the stone is dropped in and as the shaft revolves the crank and pitman moves the long end of the lever back and forward, so as to crush the stone, first on one side of feed hole and then on the other, and so crushes the stone on one side after another until it is small enough to drop through between end of lever and wedged end of feed box.

ASBESTOS LINED SECTIONAL PIPE AND BOILER COVERING.

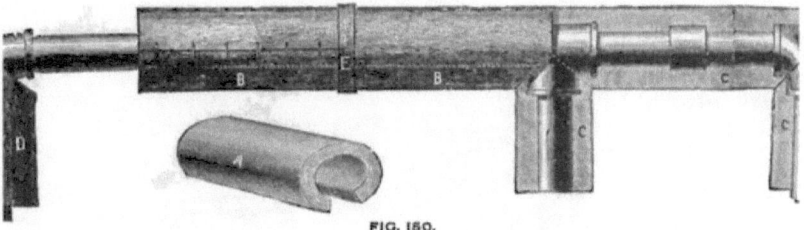

FIG. 150.

Class A is composed entirely of wool felt, and is used for water and gas pipes, and very low pressure steam pipes.

Class B is wool felt, lined inside with Asbestos. It is used principally on steam pipes, boilers, drums, etc.

Class C is wool felt, having three layers of Asbestos inside alternating with layers of wool felt. Intended for use on very high-pressure steam pipes.

It is made of compact felt, made from the pulp of woolen rags; and wool is positively the best non-conductor of heat known to science.

It is made in paper form, about an inch in thickness, is elastic and durable, and does not crack by expansion of pipes.

Being interlined with asbestos, it is indestructible at any temperature at which steam is used. Numerous instances of pipes covered with it and carrying over 160 pounds of pressure are given.

It radiates less heat, at seven-eighths of an inch in thickness, than most other coverings of twice that thickness.

From printed instructions accompanying shipments, any handy man can apply the Reed covering. When well coated with paint, it is impervious to the action of the weather whether used inside or outdoors.

It is made in sections, three feet in length, and can, if necessary, be taken off and reapplied without injury.

It does not burden down the pipes; weight of square foot, one and one-quarter pounds.

It will not fracture or loosen, and when being applied or in use makes no dirt or grit to injure machinery.

It can be applied while pipes are either hot or cold.

Gas and water pipes covered with this preserve their contents without freezing.

If desired we will do work by contract, or furnish men at fair wages and expenses.

Tests have shown that savings in fuel from 10 to 30 per cent. have been effected by this covering. Price list sent on application.

JOHN A. McCONNELL & Co.,

No. 87 Water Street, Pittsburgh, Pa.

143

ASBESTOS CEMENT PIPE COVERING WITH AIR SPACE.

FIG. 151.

This composition for lightness, efficiency and durability, is unexcelled, is elastic and will not crack from expansion of surfaces, and adheres tightly to same; is easily replaced and can be used again. The composition keys itself through the meshes of the wire lathing, where air space is adopted, and this adds to its efficiency.

AIR SPACE—FIG. 152.

The composition is made of asbestos fibre and a cementing compound, part of which is infusorial earth, widely known as fossil meal, which makes our composition an excellent non-conductor and, at the same time, very light. It is put up dry, to be mixed with water to the consistency of mortar and applied in coats by trowel on surfaces to be covered.

Our asbestos cement felting is unexcelled.

It is an excellent non-conductor; mixes readily, is easily applied, adheres tenaciously with minimum of crackage.

Write stating what you wish to cover with cement felting and we will mail you a sample of suitable grade.

For steam pipes, boilers, drums, etc., XXX,   · · · · ·   per bbl., $4 00
For hot blast pipes and all very hot surfaces, Hot Blast,   · · "   5 00

ASBESTOS AND MINERAL WOOL COMBINATION COVERING.

CLASS D—FIG. 153.

CLASS E—FIG. 154.

Both the mineral wool and the asbestos combination coverings, shown in the above cuts, combine all the advantages of the much-talked-of fire-proof coverings, together with the remarkable heat insulating qualities, everywhere recognized as being possessed by the wool felt sectional coverings.

JOHN A. McCONNELL & CO.

No. 87 Water Street, Pittsburgh, Pa.

I KNOW A FEW QUARRYMEN WHO ALWAYS COMPLAIN OF BAD ELECTRIC FUSES,

While other quarrymen never make a complaint about the same. After much investigation, I have concluded that the fault lies in the way quarrymen handle them, and to save other quarrymen trouble, I have concluded to call special attention to this subject; this I propose to do by first calling attention to the delicate construction of electric fuses or exploders, of which fig. 155 is a full-sized cut.

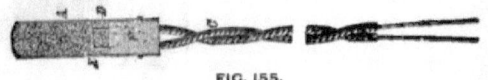

FIG. 155.

A represents the copper shell or cup, about one and a quarter inches long, and one-quarter of an inch in diameter, closed at one end and open at the other, and filled two-thirds full with explosive fulminate, B, which is very easily exploded, with the two ends, E and D, of the two copper wires, C, imbedded in the fulminate, and those two copper wires are connected to each other by a fine hair wire, E D, of platinum wire, and then the end of the cap closed by a plug of melted sulphur, F. The two wires, C, are covered with insulating covering to prevent the electricity passing from one wire to the other without going to exploder cap, and these wires are made from four to thirty feet regularly, or to any length on special order.

Now, anyone who has carefully read the above will easily understand that electric exploders are very delicately made and, of course, easily put out of order. They should be handled very carefully, for if a man catches the cap in his finger and thumb and attempts to pull a fuse out of a bundle of fifty, as they are usually put up, and should permit the wires to become tangled, as shown in fig. 156, he is very liable to injure the fuse, either by pulling the cap off and disturbing the fine hair of platinum wire, or injuring the insulation along the wires so that when the wires are laid side by side in the tamping in a hole the electricity will then jump from one wire to another and the hole missfire; then the fuse is condemned, when it has only been injured by careless handling by the quarryman, without him ever knowing he has injured it. Now, to avoid this trouble, I recommend that manufacturers of electric fuses should be required to fold every separate wire, with each fold crossed over the previous fold, as shown in fig. 158, and then packed in box, as shown in fig. 159. Then a quarryman can easily lift out as many fuses as he has holes ready to blast and take them to the holes. By catching the wires near the cap with the finger and thumb of one hand and the other end, B, with the finger and thumb of the other hand, the whole fuse will unfold itself as easily as a two-foot rule unfolds, and it can then be put into a hole with far more certainty of exploding than as now used.

Fig. 156 represents the condition I have many times found electric fuse in at a quarry where a careless man, in drawing out a few fuses from a bundle of fifty, permits the whole bundle to become tangled up as shown. Fig. 158 shows how I propose to have each separate fuse wound on a former, each layer crossed over the previous layer and each end so twisted around the layers as to hold each fuse separate, and then packed fifty in a box, as shown in fig. 159. Then when the user takes one out of the box, untwists and catches each end, as shown in fig. 157, and spreads his arms, the whole fuse will unwind, as shown in fig. 157, without injuring either cap or insulation.

FIG. 156.

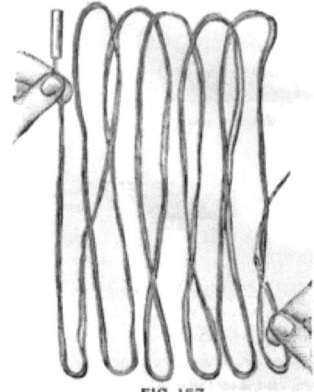

FIG. 157

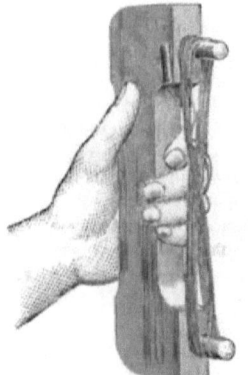

FIG. 158.

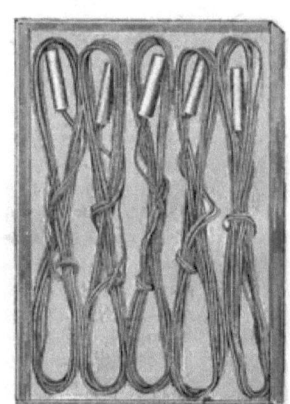

FIG. 159.

STRIPPING QUARRIES BY HYDRAULIC WASHING.

FIG. 160.

Since printing what is said about Hydraulic Washing on page 48, I have obtained the above cut and considerable information on the subject, and stop the press to give it to my readers.   Fig. 160 shows the two nozzles playing on a bank at once, and if we imagine they are five or six inches in diameter (a very common size in hydraulic washing), and throwing that volume of water with a force of from three hundred to four hundred pounds pressure at the nozzle, one can easily see that the stripping above any limestone quarry must soon give way before such a force continuously thrown against any earthen or shale bank at far less cost than by any other known means.   This force, I am assured, can easily be produced by using a duplex Hall steam pump, of suitable size and proportion, as I am informed these pumps have been repeatedly used for this purpose. Almost every quarry is located near a stream of water, where, by means of a cheap dam thrown across it, sufficient water might easily be collected in a night to run the pump part, if not all, of the next day, and do far more stripping than fifty men could do in the same time, and that merely at the cost of the wages of three men ; one to keep up steam, one to attend to the the nozzle and one to see that the water was properly conducted away after it spent its force on the bank. Another item of saving would be the washing clean of a large portion of loose stones now found imbedded in the mud and clay, and too dirty to be used for any purpose, but which, if rolled or tossed in water one or two hundred feet, would be clean enough for any use, and, although they might not be fit for fluxing purposes, yet would make excellent railroad ballast, or common road metal, as all such stone would be gotten without drilling or blasting.   I have no doubt but that in many quarries this washed stone would pay all expenses of running the pump.

LIDGERWOOD MANUFACTURING COMPANY,

99 First Avenue, Pittsburgh, Pa.

IMPROVED HOISTING ENGINES AND BOILERS, Adapted for every purpose for which Hoisting Engines are required.

MORE THAN 100 DIFFERENT STYLES

Of these Engines are built by us.

MORE THAN 7000 LIDGERWOOD ENGINES

Are in use, and all, without exception, giving perfect satisfaction.

DOUBLE CYLINDER SINGLE OR DOUBLE FRICTION DRUM HOISTING ENGINE.

Specially adapted for Contractors, Railroads and Builders' uses, Pile Driving, Excavating, Hoisting Materials, Quarries, Mines, etc.

SEND FOR OUR LARGE ILLUSTRATED CATALOGUE.

LIDGERWOOD MANUFACTURING COMPANY,

99 FIRST AVENUE, PITTSBURGH, PA.

DOUBLE DRUM REVERSIBLE LINK MOTION CLUTCH AND BRAKE HOISTING ENGINES.

Specially adapted for the Tail Rope System and Double Track Inclines, or Double Shafts in Mines.

—— —— *SEND FOR OUR LARGE ILLUSTRATED CATALOGUE.* —— ——

INDEX.

A

PAGE.

Accidents from Explosives....................... 9
Another Cause of Missfire in Electric Blasting.......... 33
Air Compressor Improve the Air in a Mine 60
Air Compressor (sectional view)....................... 70
Air Compressor, with enlarged Air Cylinder for Low
   Pressure Air....... 79
Auxiliary Valve Drill.............................. 96
Asbestos Lined Pipe and Boiler Cover............143, 144
Advantages of Compressed Air...................... 57
Allison's, Wm., Letter on Shaft Sinking.............. 62

B

Bullet-proof Magazines 8
Blasting by Electricity, Economic Value.............. 27
Blasting Reels... 28
Blasting by Electricity Compared with Fuse Blasting.. 37
Blasting and Sporting Powder..................... 44
Blasting down the walls of a Roman Catholic Church at
   Johnstown....................................... 53
Blasting the Jam at Johnstown...................... 55
Best Method of Sinking a Shaft or Running a Tunnel... 56
But when Compressed Air is Used................... 57
Bar Channeler Making Horizontal Cut................ 90
Baby Drill for Pop Holing......................... 95
Breaking Large Blocks of Granite.................... 99
Breaker, Elevator, Screen and Bin.................. 128
Breaker, Screen and Bin......................... 129
Breaker and Double-sided Bin...................... 124
Breaker with End Dump Hoist 125
Breaker Plant for Large Capacity and Fine Product..... 126
Breaker Higher Than Supply of Stone................ 127
Breaker with Screen and Return Elevator............. 128
Bomb-proof Shelters for Men....................... 12
Blasting Coal.... 24

C

Cheapest Plan for Stripping a Quarry................. 17
Cost of Electric Apparatus......................... 28
Crescent Blasting Reels........................... 29
Cause of Missfire in Electric Blasting................. 33
Contrast Between Old and New Plan of Blasting........ 36
Club Sporting Powder............................. 45
Champion Ducking Powder.......................... 46
Crack Shot Powder............................... 47
Compressed Air Can be Used........................ 57
Concentrated Piston Cold Air Compressor, Class A,
   Steam Actuated.................................. 73
Compressor with Large Cylinder for Low Air Pressure... 79
Compressor with Piston Inlet for Cold Air, Showing
   Automatic Air Pressure Cut-off................... 80
Compressor Showing Details of Automatic Air Cut-off,
   Class A or B 81

PAGE.

Compressor with Pelton Water Wheels on Main Shaft .. 82
Concentrated Piston Cold Air Compressor, Class B, Belt
   Actuated, Driven by Water Power............... 76
Concentrated Piston Cold Air Compressor, Class B,
   Driven by Belt from Water Power, Electricity or
   Stationary Engine.............................. 77
Compound High Pressure Air Compressor, for air pres-
   sure up to 800 pounds to one inch................ 78
Compressor Showing Enlarged View of Automatic Air
   Pressure Regulator.............................. 81
Coal Miners' Drills, Scrapers, etc................... 109
Cambria Iron Works Plant......................... 122
Cowles' Patent, preventing accident from over-hoist of
   cage..................................134, 135, 136
Cotton Fuse.. 25

D

Dynamite Thawing, Plan of Room................... 10
Dynamite Thawing, Kettles......................... 13
Does Squibbing Holes Pay?.......................... 15
Dynamite as an Explosive.......................... 19
Demand for High Explosives........................ 19
Dynamite May be Used Laid on Rock................. 20
Dynamite is Invaluable for Stump Blasting........... 20
Dynamite, How to Use............................. 21
Dynamite Made Perfectly Safe...................... 20
Directions for Using Crescent Reel.................. 29
Directions for Charging Holes to Fire by Electricity.... 30
Directions for Blasting by Electricity................ 30
Duplex Pumps Should Be Used in Shaft Sinking...... 59
Details of Construction of Air Compressors........... 71
Duplex Compound Corliss Air Compressor............ 73
Duplex Compound Air Compressor, ground plan....... 75
Dynamite, Several Grades of....................... 58
Directions for Using Kirk's Water Injector.......... 86
Descriptive Table of Rock Drills................... 105
Directions for Operating Rock Drills................ 106
Directions for Setting Up and Care of Stone Breakers... 114
Ditman's Patent Hand-power Rock Drill.............. 140
Dam for Hydraulic Washing......................... 18
Double Tape Fuse................................. 25
Dynamite Explosion............................... 55

E

Elevation of Thawing House........................ 11
Electric Exploders, How to Fasten Exploder to Cartridge 23
Electric Blasting in Shaft Sinking................60, 61
Electric Blasting Compared with Fuse Blasting....... 41
Elevation of Air Compressor, Air Receiver and Boiler.. 74
Engineers' Attention Called to the Importance of Com-
   pressed Air 70
Electric Exploders, Perfection in Machinery for Making. 33
Electric Blasting................................. 38
Elevators 118
Electric Fuses, How to Handle..................145, 146

INDEX.

F

	PAGE.
Front Elevation Thawing House.	12
Firing by Fuse	25
Frick & Co.'s Letter Recommending Compressed Air for Running Pumps	74
Fuse Blasting	40
Ford's Paten Tunnel Feed and Water Heater.	66
Flat Hole Work with Power Drill	98
Forster Rock Breaker	142
Fastening Electric Exploder in a Dynamite Cartridge	23

G

	PAGE.
Great Need of Better Laws	7
Great Danger in Squibbing Holes	15
Great Perfection of Making Electric Exploders	33
Grim Patent Coal and Clay Boring Machines	107
Gates Rock and Ore Breaker	112
Gutta-Percha Fuse	25

H

	PAGE.
Heavy Blasting	16
How to Fire Blasts in Coal, Rock and Stumps	24
How to Fasten Fuse to Cartridge	26, 27
How to Fasten Electric Exploder to Dynamite Cartridge	23-26
How to Connect the Wires for Blasting	31, 32, 33
How to Blast Salamander	43
How to Use a Shot Gun	45
How to Charge Holes in a Salamander	44
How Blasting in the Johnstown Jam was Done	54
Horizontal Tubular Boilers, Full Front	67
Hydraulic Washing Best Plan to Strip a Quarry	17, 147
Horizontal Steam Engine	67
Hose and Couplings	101
Hand Power Drill	140
Hoisting Engines	148, 149
Hemp Fuse	25

I

	PAGE.
Introduction	3
Importance of Blasting Explosives	3
Insurance of Explosives	7
Importance of Re-examining Wires	32
Incrustation in Steam Boilers	64
I do not Recommend Fancy Boilers	66
Inlet Air Piston to Air Compressor	70
Important Features of the Auxiliary Valve Drill	94
Improved Hoisting and Conveying	129, 130, 131, 132, 133
Improved Plan of Packing Fuses	145, 146

J

	PAGE.
Johnstown Jam, 350 feet wide by 850 feet long	50
Johnstown Jam Above Stone Bridge	51, 52, 53, 54, 55

K

	PAGE.
Kitchen Range Explodes	6
Kirk's Water Injector for Power Drills	85

L

	PAGE.
Larger Hole at Bottom	15
Leading Wires	28
Letter from J. K. Taggart	58
Letter from William Allison	62
Letter from H. C. Frick & Co.	74
Letter from H. L. Schweyer	93
Ledgerwood Mfg. Co., ad	148, 149

M

	PAGE.
Majendie Letter, Col. V. D.	5
Making Small Blasts	16
Money Lost in Loading Stone	17
Magnetic Machine for Blasting	28
My Experience with Dynamite at Johnstown in 1889	48, 49, 50, 51, 52, 53, 54, 55
Machinery Department of Quarry Work	63
Missfire in Blasting, Cause of	33
Modes of Firing, Result of	36
Method of Shaft Sinking	56
Mining or Tunnelling Plant, showing boilers, pump, compressor, etc.	83
Morna Flexible Steam Joint, to take the place of steam and air hose	100
Matches Worse than Dynamite	6
McConnell, Jno. A. & Co., ad.	143, 144

N

	PAGE.
New Plan of Heavy Blasts	35
New Mode of Blasting Dimension Stone	42
No Hang-fire Shots	39

O

	PAGE.
Old Plan of Heavy Blasts	34
Opening Powder Kegs	42
Oil and Tools for Power Drills	87
Oiling Stone Breaker	116

P

	PAGE.
Powder Magazines	7, 141, 142
Plenty of Holes Ready for Blasting	14
Paying Men by the Day	17
Powder Spoon	26
Pumping Power	57
Pumps	58
Plan of Air Compressors, Boiler, etc	78
Producing First Point of Rupture at Bottom of Hole	38
Power Rock Drills	84
Price of Rock Drills	102
Price of Complete Plant of Mining Machinery	103

INDEX.

P

PAGE.
Power Drills Should Be Used in Fire Clay Mines.......108
Plans for Changing Parts of Breaker.................117
Portable Stone Breaker..............................119
Pipe Covering.................................143, 144
Plan of Dynamite Thawing Room.....................10

Q

Quarrying with Bar Channeler.......................92
Quarrymen Complain of Bad Electric Fuses..........145

R

Rules to be Observed in Magazines..................9
Rock and Ore Breaker.............................110
Result of the Two Modes of Firing....36, 37, 38, 39, 40, 41
Reckless Way of Opening Powder Kegs...............42
Rifle Powder......................................46
Reason why the water struck Johnstown with so much
    force...49
Running a Tunnel with Power Drills and Compressed Air. 90
Rock Drill Hose and Couplings....................101
Renewing Hopper Liners...........................113
Revolving Screens................................118
Road Roller, Best Pattern........................129
Rope Hoist and Conveying Apparatus...............129
Rock Crusher and Pulverizer Combined......137, 138, 139
Running a Heading in a Tunnel.....................85
Rock Drill and Ore Breaker.................110 to 128

S

Stripping a Quarry, Cheapest Plan.............17, 146
Stump Blasting Made Easy and Safe.................21
Stump Blasting, Large or Hollow...................22
Stump Blasting, Fresh Cut, etc....................22
Squib Firing......................................24
Smokeless Powder..................................47
Shaft Sinking............56, 57, 58, 59, 60, 61, 62
Steam Should be Used Through an Air Compressor.....57
Steam Boilers.....................................63
Stationary Tubular Steam Boilers..................64
Stationary Tubular Steam-setting Boilers..........66
Stationary Tubular Steam Table of Specifications.......65

P

PAGE.
Stationary Horizontal Steam Engine................67
Stationary Vertical Steam Engine..................68
Shot Gun, Use of..................................45
Single Quarry Bar.................................89
Sinking a Large Shaft with Power Drills...........91
Starting a Quarry with Power Drills...............97
Specifications and Prices of Drill Steels.........104
Size and Weight of Column.......................105
Starting Stone Breaker...........................116
Stone Breaker Mounted on Wheels..................119
Steam Engine for Hoisting and Conveying Apparatus.130, 131
Special Engine for Hoisting and Conveying.......130, 131
Single Tape Fuse..................................25
Sectional View of Air Cylinder....................70
Schweyer, H. I., letter on Advantage of Quarry Machin-
    ery...93
Steam Hose, Prices, etc..........................101
Stripping Quarries by Hydraulic Washing..........147
Sinking Shaft.....................................62

T

Thawing Dynamite in an Improper Way....9, 10, 11, 12, 13
Thawing Dynamite, Kettles.........................13
Thawing Dynamite, Room............................10
The Whole Mass Caught Fire........................50
The Great Johnstown Jam...........................50
Then Use Electric Blasting........................58

U

Use Dynamite After Trying Several Grades..........58
Use Ford's Patent Water Heater in all Steam Boilers....67
Use of Compressed Air.............................57

V

Very Best Machinery for the Work..................63
Vertical Steam Engines............................68

W

Wrong Treatment Imposed on Explosives..............4
Worse Than Dynamite................................6
Waste Force in Handling Minerals..................14
Work for which Drill is Best Suited...............87